# Tissue Proteomics

METHODS IN MOLECULAR BIOLOGY™

# Tissue Proteomics

## Pathways, Biomarkers, and Drug Discovery

Edited by

## Brian C.-S. Liu

*Brigham and Women's Hospital,*
*Harvard Medical School,*
*Boston, MA*

and

## Joshua R. Ehrlich

*Assistant Editor*
*Brigham and Women's Hospital,*
*Harvard Medical School,*
*Boston, MA*

*Editors*
Brian C.-S. Liu
Brigham and Women's Hospital
Harvard Medical School
Boston, MA
USA

Joshua R. Ehrlich
Assistant Editor
Brigham and Women's Hospital
Harvard Medical School
Boston, MA
USA

*Series Editor*
John M. Walker
School of Life Sciences
University of Hertfordshire
College LaneHatfield, Herts. AL10 9AB
Hatfield, Hertfordshire AL10 9AB
UK

ISBN: 978-1-58829-679-5     e-ISBN: 978-1-60327-047-2

Library of Congress Control Number: 2007939568



Cover illustration: Fig. 2A of Chapter 5 by Adrianna S. Rodriguez et al.

Printed on acid-free paper

9 8 7 6 5 4 3 2 1

springer.com

# Preface

With the sequencing of the human genome complete, the field of proteomics, still in its early stages, has become an important and informative field of biomedical research. Many of the protocols and techniques commonly employed by proteomics researchers have been refined over the past several years, while at the same time, new and innovative methods have also been developed. Significantly, many of the protocols used in the study of the human proteome have been developed to examine proteins derived specifically from cell lines, body fluids or human tissues. As a consequence, not all protocols are compatible with the full-range of protein sources commonly used in the laboratory. The current volume on Tissue Proteomics seeks to bring together a number of useful and innovative protocols developed particularly for the proteomic profiling of human tissues.

The first two protocols presented in this work focus on high-throughput gel-based techniques. Chapter 1 describes 2-D DIGE, a method used to examine the entire proteome within discreet pI fractions, whereas Chapter 2 details several complementary techniques for the specific analysis of glycoproteins found in tissue samples. The next chapters, Chapters 3 and 4, focus on SELDI-MS and MALDI-TOF, two techniques that make use of mass spectrometry for the characterization and identification of proteins. These techniques may be used alone or in combination with other proteomic methods, such gel-based assays, where mass spectrometry is useful for additional characterization and identification of proteins.

The fifth chapter in this volume is unique in that it presents a very useful protocol, however, one that must be used in conjunction with other methods. Automated laser capture microdissection, which is used to capture homogeneous cell populations from crude tissue samples may be used in concert with a number of the other platforms presented in this volume. This technique should be considered a valuable tool for researchers in tissue proteomics, especially for the production of custom arrays.

The chapters that follow, Chapters 6–10, focus on the production and/or use of microarrays for the purpose of profiling protein expression patterns in tissue samples. Chapters 6 and 7 both describe the construction of custom arrays using tissue-derived antigens as arrayed features that may be assayed and quantified. Chapter 8 then presents a reverse-phase array platform that, due to its high specificity, is an effective tool for measuring specific protein and protein-modification levels. In Chapter 9, a commercially designed monoclonal antibody array is presented that can be used to capture and measure expression of 500 unique proteins involved in many important cellular processes. Finally, Chapter 10 takes advantage of the specific affinity of SH2 domains for tyrosine phosphorylations in order to construct an array that can be used to assay levels of these important molecular modifications in tissue samples.

The focus of Chapters 11 and 12 is on measuring autoantibody expression for tissue-based proteomic research. Chapter 11 depicts the use of arrayed transcription factors to identify serum autoantibodies that may be used as biomarkers and reporters of aberrantly regulated transcription factors. Similarly, Chapter 12 presents a novel and useful method for detecting autoantibody expression to a wide variety of native intracellular antigens that are immobilized on a monoclonal antibody microarray.

Lastly, we have included an essential final chapter that discusses the standardization of methods for sample collection, processing and storage of human tissues. All future research in tissue proteomics will depend upon the implementation of effective standards that allow results to be validated, compared and built-upon between institutions.

Together, the 13 protocols presented in this volume represent a number of the newest and most innovative methods in tissue proteomics. It is our desire that these protocols serve as a helpful resource and guide to investigators, while aiding future work in the areas of biomarker discovery, the profiling of therapeutic responses, and the new era of theranostics and personalized medicine.

This book could not have been completed without the hard work and contribution of a large number of individuals. The editors would like to thank the many people involved in this process, especially the authors who have submitted these new and innovative protocols. In addition, the editors would like to thank Robert J. Caiazzo Jr. for his assistance in the preparation and completion of this book.

# Contents

# Contributors


MARGARITA APONTE • *Laboratory of Gynecologic Oncology, Department of Obstetrics, Gynecology, and Reproductive Biology, Brigham and Women's Hospital, Harvard Medical School, Dana-Farber Cancer Center, Boston, MA*

ROBERT J. CAIAZZO JR. • *Molecular Urology Laboratory, Division of Urology, Brigham and Women's Hospital, Harvard Medical School, Boston, MA*

VALERIE S. CALVERT • *Center for Applied Proteomics and Molecular Medicine, George Mason University, Manassas, VA*

GRIGORIY S. CHAGA • *Clontech Laboratories, Mountain View, CA*

DANIEL W. CRAMER • *Department of Obstetrics and Gynecology, Brigham and Women's Hospital, Harvard Medical School, Boston, MA*

RAJIV DHIR • *UPMC Shadyside-Presbyterian Hospital, Department of Pathology, Pittsburgh, PA*

JOSHUA R. EHRLICH • *Molecular Urology Laboratory, Division of Urology, Brigham and Women's Hospital, Harvard Medical School, Boston, MA*

BENJAMIN H. ESPINA • *National Cancer Institute, Center for Cancer Research, Laboratory of Pathology, Bethesda, MD*

VIRGINIA ESPINA • *National Cancer Institute, Center for Cancer Research, Laboratory of Pathology, Bethesda, MD and Center for Applied Proteomics and Molecular Medicine, George Mason University, Manassas, VA*

AUDREY GAGNON • *Laboratory of Gynecologic Oncology, Department of Obstetrics, Gynecology, and Reproductive Biology, Brigham and Women's Hospital, Harvard Medical School, Dana-Farber Cancer Center, Boston, MA*

STEPHANIE HAN • *Panomics, Inc., Dumbarton Cr., Freemont, CA*

STEPHEN M. HEWITT • *Tissue Array Research Program, Laboratory of Pathology, Center for Cancer Research, NCI, and Renal Diagnostics and Therapeutics Unit, NIDDK, NIH, Bethesda, MD*

XIN JIANG • *Panomics, Inc., 6519 Dumbarton Cr., Freemont, CA*

SAU-MEI LEUNG • *Clinical Proteomics, Bruker Daltonics, Inc., Billerica, MA*

XIANQIANG LI • *Panomics, Inc., Freemont, CA*

LANCE A. LIOTTA • *National Cancer Institute, Center for Cancer Research, Laboratory of Pathology, Bethesda, MD and Center for Applied Proteomics and Molecular Medicine, George Mason University, Manassas, VA*

BRIAN C.-S. LIU • *Molecular Urology Laboratory, Division of Urology, Brigham and Women's Hospital, Harvard Medical School, Boston, MA*

STACEY M. LOYLAND • *GE Healthcare, 800 Centennial Ave., Piscataway, NJ*

SHU-KAY NG • *Department of Mathematics, University of Queensland, Brisbane, Australia*

SHU-WING NG • *Department of Obstetrics and Gynecology, Brigham and Women's Hospital, Harvard Medical School, Boston, MA*

EMANUEL F. PETRICOIN III • *Center for Applied Proteomics and Molecular Medicine, George Mason University, Manassas, VA*

REBECCA L. PITTS • *Clinical Proteomics, Bruker Daltonics, Inc., Billerica, MA*

ADRIANNA S. RODRIGUEZ • *National Cancer Institute, Center for Cancer Research, Laboratory of Pathology, Bethesda, MD*

LESILE ROTH • *Panomics, Inc., 6519 Dumbarton Cr., Freemont, CA*

CHRISTINE R. ROZANAS • *GE Healthcare, 800 Centennial Ave., Piscataway, NJ*

QIAN SHI • *Laboratory of Gynecologic Oncology, Department of Obstetrics, Gynecology, and Reproductive Biology, Brigham and Women's Hospital, Harvard Medical School, Dana-Farber Cancer Center, Boston, MA.*

INJAE SHIN • *Department of Chemistry, Yonsei University, Seoul, 120–749, Korea*

ROBERT A. STAR • *Tissue Array Research Program, Laboratory of Pathology, Center for Cancer Research, NCI, and Renal Diagnostics and Therapeutics Unit, NIDDK, NIH, Bethesda, MD*

LIANGDAN TANG • *Department of Obstetrics and Gynecology, Brigham and Women's Hospital, Harvard Medical School, Boston, MA*

OLIVER W. TASSINARI • *Molecular Urology Laboratory, Division of Urology, Brigham and Women's Hospital, Harvard Medical School, Boston, MA*

DEAN TROYER • *Department of Pathology, University of Texas Health Science Center, San Antonio, TX*

JULIA WULFKUHLE • *Center for Applied Proteomics and Molecular Medicine, George Mason University, Manassas, VA*

BIN YE • *Laboratory of Gynecologic Oncology, Department of Obstetrics, Gynecology, and Reproductive Biology, Brigham and Women's Hospital, Harvard Medical School, Dana-Farber Cancer Center, Boston, MA*

ALINA D. ZAMFIR • *Institute fuer Medizinische Physik und Biophysik, D-48149 Muenster, Germany*

# 1

# Capabilities Using 2-D DIGE in Proteomics Research

*The New Gold Standard for 2-D Gel Electrophoresis*

**Christine R. Rozanas and Stacey M. Loyland**

## Summary

The use of two-dimensional gel electrophoresis for differential analysis in proteomics was revolutionized by the introduction of 2-D fluorescence difference gel electrophoresis (2-D DIGE). This fluorescence-based technique allows the use of multiplexed samples and an internal standard that virtually eliminates gel-to-gel variability, resulting in increased confidence that differences found between samples are due to real induced changes, rather than inherent biological variation or experimental variability. 2-D DIGE has quickly become the "gold standard" for gel-based proteomics for studying tissues, as well as cell culture and bodily fluids such as serum or plasma.

This chapter will describe the basic 2-D DIGE technique using minimal labeling, image acquisition using high-quality fluorescence scanners, and software capable of analyzing the multiplexed images and normalizing the data using the internal standard.

**Key Words:** Difference Gel Electrophoresis; multiplexing; fluorescent labeling; two-dimensional gel electrophoresis; DeCyder; DIGE.

## 1. Introduction

The study of proteomics relies on technologies capable of separating and resolving hundreds or thousands of proteins to a degree sufficient to allow subsequent analysis. Further methods of analysis on the resolved proteins include determining relative protein abundance changes because of environmental or genetic changes within the cell or organism, followed by identification by mass spectrometry (MS), or confirmatory analysis [(Western blots or enzyme-linked immunosorbent assay (ELISA)]. Two-dimensional gel

From: *Methods in Molecular Biology, vol. 441: Tissue Proteomics: Pathways, Biomarkers, and Drug Discovery*
Edited by: B.C.-S. Liu © Humana Press, Totowa, NJ

electrophoresis (2-DE) has taken a prominent role in proteomics because of its unrivaled ability to separate thousands of proteins with high resolution in an attractive visual presentation, which is easily coupled to downstream analysis as described above. However, the ability to perform quantitation on 2-D gels has been hampered by the high variability introduced by the many steps and processes required to produce the gels. Until recently, quantitation on 2-D gels required running high numbers of technical replicates of each sample, as well as biological replicates, resulting in variable ranges of expression levels, and little confidence in changes below a twofold difference.

The use of 2-D gels, particularly for quantitative proteomics or differential analysis, has been revolutionized with the development of 2-D fluorescence difference gel electrophoresis (2-D DIGE). 2-D DIGE was first developed in 1997 by John Minden *(1)* and commercialized by Amersham Biosciences (now GE Healthcare), Piscataway, NJ. It was a simple idea based on two concepts: running more than one sample on a gel at one time to address the complexity of proteomics research and using an internal standard to address the issue of gel-to-gel variability. The technique is based on the use of spectrally resolved fluorescent cyanine dyes (CyDye™ DIGE fluors) that can label proteins and provide a wide linear dynamic range (four orders) of quantitation through fluorescence, rather than by the traditional colorimetric methods such as Coomassie™ blue or silver staining that have limited dynamic ranges. GE Healthcare now offers two approaches to labeling, one using lysine labeling with the minimal dyes for general sample use or, alternatively, using cysteine labeling with the scarce sample-labeling kit for samples in limited quantities, such as tissue samples collected by laser capture microdissection *(2,3)*. The 2-D DIGE approach has been used to analyze tissues from adipose *(4)*, brain *(5–8)*, breast *(9,10)*, colon *(11,12)*, endometrium *(13)*, esophagus *(3)*, gastric tumors *(14)*, heart *(15,16)*, intestines *(2)*, liver *(17–20)*, lung *(21,22)*, muscle *(23)*, pancreas *(22)*, visual cortex *(24,25)*, and numerous other sample types such as cell cultures and serum. This chapter will focus on the three key components of the technology: *(1)* the dyes themselves, focusing on the use of minimal dyes, *(2)* the scanner, and *(3)* the software that can be used to take advantage of this approach.

In the 2-D DIGE method (*see* **Fig. 1**), two or more samples to be compared are labeled with the appropriate CyDye DIGE fluors and then combined after labeling. The mixture is separated by isoelectric focusing (IEF) and second dimension sodium dodecyl sulfate–polyacrylamide gel electrophoresis (SDS–PAGE), using standard gel equipment (*see* **Note 1**). The dyes have been designed to carry a +1 charge that replaces the +1 charge lost on the lysine epsilon amino group because of labeling. Thus, the pI of the labeled protein does not differ from the unlabeled protein *(1)*. In addition, the dyes have been

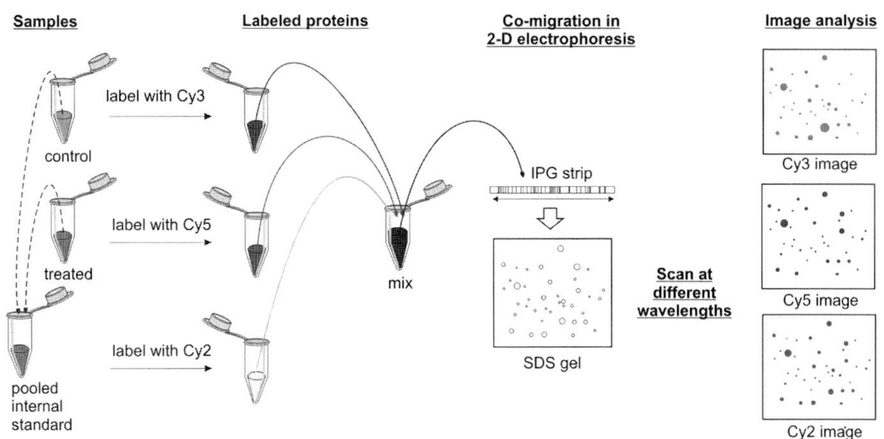



Fig. 1. Difference gel electrophoresis (DIGE) workflow. An aliquot of each sample is mixed to create the internal standard. All samples and the standard are labeled, and then combined before running the 2-D gel. Image acquisition and analysis complete the workflow. Obtained with permission from Westermeier R, Scheibe B. Difference gel electrophoresis based on lys/cys tagging. In Posch A. Ed. *Sample Preparation for 2D PAGE. Methods in Molecular Biology*, Humana Press, Totowa, NJ (2007).

designed to have very similar molecular weights (MW). These two factors are key, as the same labeled protein from different samples will migrate to the same position on the 2-D gel where the labeled protein can be imaged for the presence of each dye, indicating the abundance of that protein in each original sample. The description "minimal labeling" derives from the ratio used to label the protein with dye. The dye : protein ratio is deliberately kept low, allowing only a small portion of lysines to be labeled. At the ratio used, sensitivity is equal to or better than silver staining and reduces the chance for multiple labeling, which could cause spot trains (multiple spots for the same protein) in the gel and affect the ability to perform accurate quantitation on the protein. The presence of a single attached dye adds approximately 450 kDa to the labeled protein. This means the labeled and unlabeled populations of each spot will migrate slightly differently from each other. The effect is negligible for high-MW proteins but is more marked for lower MW proteins. For quantitation, this offset on the gel does not present a problem, because the labeled versions all migrate the same regardless of protein size. However, if such a spot is to be picked for MS identification, the gel should be post-stained, generally with a fluorescent stain such as Deep Purple™ or SYPRO™ Ruby. This will locate the gel position with the highest protein concentration corresponding to the protein of interest and allow correct positioning of the spot picker coordinates.

The ability to multiplex samples within the same gel allowed the introduction of the second key concept of 2-D DIGE: the internal standard. The use of the internal standard allows a number of improvements over traditional 2-D gels including reduced gel-to-gel variation to ensure accurate quantitation through normalization across gels; the ability to see changes that would be missed if not using an internal standard *(11)*; the ability to see changes well below the twofold threshold, opening up the possibility of finding differences in protein expression not previously discovered, particularly for low abundant proteins, as related to a particular disease state or drug treatment; reducing the chances of false positives and false conclusions about the biological implications of the protein expression changes seen (*see* **Fig. 2**); and, the ability to compare multiple samples more easily and rapidly across gels through the link provided by the internal standard. The importance of the internal standard was confirmed by comparing the results of a spiked sample analyzed with and without including the internal standard *(26)*. Further refinements include the use of dye swapping between samples (*see* **Table 1**) and randomizing the samples within the experiment (e.g., Control 1 and Treatment 2 might be paired on one gel, while Treatment 3 and Control 2 on a different gel, and so on), reducing any potential bias arising from the labeling or pairing of samples in the experiment. The internal standard is created by pooling aliquots from each individual sample in the experiment into one tube (*see* **Table 2**). The pooled standard sample is labeled with Cy™2 in

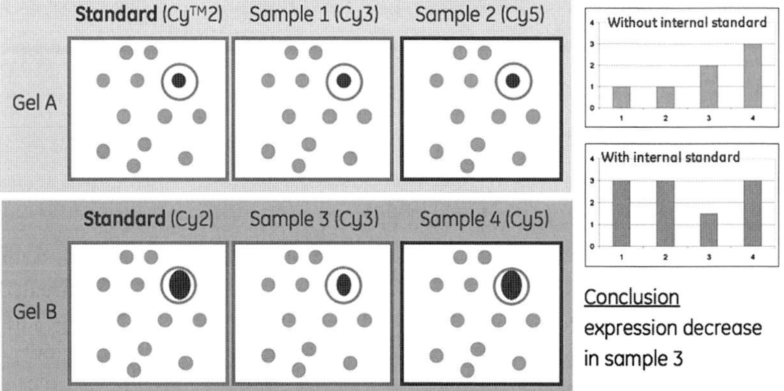


Fig. 2. The internal standard concept can be illustrated as shown using two gels with four samples. The circled spot appears to be increased in samples 3 and 4 as compared with samples 1 and 2 based on volume alone. However, after normalization with the internal standard on both gels, it is obvious that sample 4 is equivalent in abundance to samples 1 and 2, whereas sample 3 shows an actual decrease. A false conclusion would have been drawn if not using an internal standard.

**Table 1**
**Example of an Experimental Design for Four Samples**
***(1–4)* with Four Biological Replicates Each (a–d)**

| Gel | Cy2 | Cy3 | Cy5 |
|---|---|---|---|
| 1 | Internal Standard | **1a** | 3d |
| 2 | Internal Standard | 3c | 4a |
| 3 | Internal Standard | 2c | 4c |
| 4 | Internal Standard | 4d | **1b** |
| 5 | Internal Standard | 3a | 2a |
| 6 | Internal Standard | **1c** | 2b |
| 7 | Internal Standard | 4b | 3b |
| 8 | Internal Standard | 2d | **1d** |

Each sample is reverse labeled on two replicates, and all samples are randomly paired with every other sample. Sample 1 is in boldface to demonstrate this design.

the minimal labeling method, and then a portion of this pool is added to a pair of samples labeled with Cy™3 or Cy™5 so that a Cy2-standard is present on each gel. This pooled standard represents a global standard for the experiment, containing all possible proteins that are present in any of the samples within the experiment, and provides a standard for each individual protein spot on every gel, which can be linked and normalized across all the gels within the experiment.

The samples are then separated by 2-DE following standard protocols. After the sample separation is completed, the gels must be imaged to view the results of the gel run. Either a high-performance laser scanner or charge-coupled device

**Table 2**
**Calculation of an Internal Standard**

| | Cy2-Internal Standard | Cy3 | Cy5 |
|---|---|---|---|
| Gel 1 | 50 µg Internal Standard | 50 µg Sample 1, treated | 50 µg Sample 2, untreated |
| Gel 2 | 50 µg Internal Standard | 50 µg Sample 3, untreated | 50 µg Sample 4, treated |

Internal Standard: 25 µg Sample 1 + 25 µg Sample 2 + 25 µg Sample 3 + 25 µg Sample 4 = 100 µg. Split the labeled internal standard sample in half to include on the two gels.

For an experiment containing four samples, 50 µg of each sample is required. In addition, for two gels, 100 µg of internal standard is needed (50 µg per gel). From four samples, 25 µg from each sample is required to form the internal standard (25 µg Sample 1 + 25 µg Sample 2 + 25 µg Sample 3 + 25 µg Sample 4 = 100 µg). Split the labeled internal standard sample in half to include on the two gels. Thus, each sample must contribute 75 µg (50 µg + 25 µg for the standard) to this experiment.

(CCD) camera-based imager, capable of distinctly imaging each of the CyDye DIGE fluors, can be used. High-quality fluorescent images must be acquired from each labeled sample to maximize the ability to locate all differentially expressed proteins, including low abundant proteins of interest. The emission and excitation filters, or the laser settings used for each dye, should be narrow band to reduce any chance of cross-talk between the labeled samples. For laser scanners, a blue laser is used for Cy2; a green laser for Cy3; and a red laser for Cy5. *See* **Table 3** for excitation and emission settings. The acquisition settings should be set below saturation levels, as saturation would compromise the quantitation of those spots. Having a scanner or imager that can link three concurrent scans ensures that the three images acquired are identical and overlay properly for proper image analysis in the next step.

The final key component to 2-D DIGE is the software. The introduction of the internal standard required a new type of software to analyze the multiplexed images from each gel, as those images were directly super-imposable upon each other. When 2-D DIGE was introduced, no software was available that could take advantage of the multiplexed images or the internal standard. Amersham Biosciences (now GE Healthcare) introduced a software platform, DeCyder™ 2-D Differential Analysis software (DeCyder 2-D), designed to take full advantage of the internal standard and multiplexed images, as well as provide high-throughput batch processing, advanced statistical tools, and integration with spot picking and MS. Recently, GE Healthcare introduced ImageMaster™ Platinum DIGE-enabled software, for lower throughput laboratories, with the same spot detection algorithm that is found in DeCyder 2-D. Further detailed descriptions of DeCyder 2-D can be found in the *Methods* section of this chapter.

The use of 2-D DIGE with 2-D gels has provided the means necessary to perform differential analysis on complex samples by reducing the gel-to-gel variation, therefore allowing detection of small changes with high reproducibility. The DIGE labeling and experimental design, combined with a

**Table 3**
**Scanner Settings on Typhoon for CyDye and Deep Purple**

| Dye | Color of Dye | Emission Filter (nm) | Laser |
|---|---|---|---|
| Cy2 | Yellow | 520 BP40 | Blue2 (488) |
| Cy3 | Red | 580 BP30 | Green (532) |
| Cy5 | Blue | 670 BP30 | Red (633) |
| Deep Purple | Purple | 560 LP | Green (532) |

Filter parameters may vary depending on instrument but should be narrow band where possible for best acquisition results.

high-quality scanner and software, ensures the best information can be obtained with the highest confidence.

## 2. Materials

### 2.1. Lysis and Labeling

1. CyDye DIGE Fluor minimal kit, 5 nmol each of Cy2, Cy3, and Cy5 (Cat. No. 25-8010-65, GE Healthcare). CyDye dyes are also available individually in 5, 10, and 25 nmol packages.
2. Dimethyl formamide (DMF) (Cat. No. D-4551 Sigma Aldrich, St. Louis, MO, USA) (*see* **Note 2**).
3. 10 mM lysine (L-Lysine, Cat. No. L-5626, Sigma Aldrich).
4. PlusOne™ Urea (Cat. No. 17-1319-01, GE Healthcare).
5. PlusOne Tris (Cat. No. 17-1321-01, GE Healthcare).
6. PlusOne CHAPS (Cat. No. 17-1314-01, GE Healthcare).
7. Thiourea (Cat. No. RPN 6301, GE Healthcare).
8. pH Test Strips range 4.5–10 (Cat. No. P4536, Sigma Aldrich).
9. Lysis buffer (7 M urea, 2 M thiourea, 4% CHAPS, 30 mM Tris, pH 8.5).
10. 2× sample buffer (7 M urea, 2 M thiourea, 2% CHAPS, 130 mM DTT).
11. Microcentrifuge tubes.
12. 18 Mega-ohm water.
13. Ettan™ 2-D Quant Kit (Cat. No. 80-6483-56, GE Healthcare).
14. Ettan 2-D Clean-up kit (Cat. No. 80-6484-51, GE Healthcare).
15. Crushed ice and ice bucket.
16. Pipettors capable of microliter volumes.
17. Latex or other laboratory disposable powder-free gloves.

### 2.2. SDS–PAGE Gel Preparation

Standard gel pouring reagents are used. In addition, the following are needed for 2-D DIGE gels.

1. Low Fluorescence Glass Plates (Cat. No. 80-6475-58 for DALT size, 80-6442-14 for SE600 size, GE Healthcare) (*see* **Note 1**).
2. PlusOne Bind-Silane (Cat. No. 17-1330-01, GE Healthcare) (*see* **Note 3**).

### 2.3. Scanning, Fixing, and Post Staining

1. Crew Wiper lint free wipes (Cat. No. 21908-205, VWR Scientific, West Chester, PA, USA).
2. Deep Purple Total Protein Stain (Cat. No. RPN 6306, GE Healthcare).
3. Glacial Acetic Acid (Cat. No. EM-AX0073, VWR Scientific).
4. Ethanol (Cat. No. BJAH090-4, VWR Scientific) (*see* **Note 2**).
5. Rocking table.
6. Plastic trays.

# 3. Methods

## 3.1. Sample Preparation for Labeling

Tissue preparation techniques include sonication *(11)*, homogenization *(15,18)*, freezing and pulverizing *(16)*, sectioning with a cryostat *(10,11)*, and laser capture microdissection *(2,3,8,14)*. Solubilization of the proteins from the tissue should be done with a lysis buffer that is compatible with labeling, such as 7 M urea, 2 M thiourea, 30 mM Tris, and 4% CHAPS (pH 8.5). The lysis solution should not contain any primary amines, such as ampholytes or thiols (e.g., DTT) as these will compete with the proteins during labeling, resulting in fewer dye-labeled proteins. If alternative reagents are to be used, refer to the compatibility list provided with the dyes, or their compatibility with labeling must be tested *(27)*. If certain reagents are required for solubilization, they can be removed through buffer exchange or dialysis before labeling. After the sample is solubilized, the pH should be adjusted and protein concentration determined (*see* **Notes 4** and **5**). Sample concentration should be between 1 and 10 mg/ml for best labeling results.

## 3.2. Preparation of the Internal Standard

The internal standard is created by pooling an aliquot of all biological samples in the experiment and labeling the pool with one of the CyDye DIGE Fluor dyes (usually Cy2 when using CyDye DIGE Fluor minimal dyes). Sufficient material is required to add equal amounts to each gel required for the experiment, typically 50 µg per standard per gel. See **Table 2** for an example.

## 3.3. CyDye Preparation

The CyDye DIGE Fluor minimal dyes for labeling are prepared by reconstitution of the solid powder into a stock solution, followed by further dilution to a working solution. The stock solution is prepared as follows:

1. Take the CyDye DIGE Fluor dye from the –20 °C freezer, spinning briefly to ensure that the powder is at the bottom of the tube, and leaving to warm for 5 min at room temperature without opening.
2. The dyes are reconstituted with high-quality anhydrous DMF by adding the volume described in the specification sheet supplied with the dye to give a final concentration of 1 nmol/µl or 1 mM. For example, 25 µl DMF is added to 25 nmol of dye (1 mM).
3. Once the DMF is added to each vial of dye, replace the cap on the microfuge tube containing the dye, and vortex vigorously for 30 s to dissolve the dye.
4. Centrifuge the microfuge tube for 30 s at 12 000 × *g* in a microcentrifuge. After reconstitution in DMF, the dye will give a deep color; Cy2-yellow, Cy3-red, and

Cy5-blue. Once reconstituted, the minimal dye stock solution is stable for 2 months at –20 °C or until the expiry date on the container, whichever is sooner.

A working solution is prepared from the stock solution and used for labeling. To make the recommended 400 pmol/µl working solution:

5. Briefly spin down the dye stock solution in a microcentrifuge. Add 2 µl dye stock solution to 3 µl DMF.
6. Ensure all dye is removed from the pipette tip by pipetting the working dye solution up and down several times.

The working dye solution is stable for 1 week at –20 °C but is generally prepared fresh for each use. The dye solutions should be returned to the freezer as soon as possible after labeling is performed.

### 3.4. Sample Labeling

The dye-labeling reaction is designed to be simple and should take about 45 min to complete. The standard protocol recommends that 400 pmol of dye be used to label 50 µg of protein. If labeling more than 50 µg of protein then the dye : protein ratio must be maintained for all samples in the same experiment. Other dye : protein ratios can be used but must be optimized for the sample by testing the labeling using a 1-D SDS–PAGE gel *(27)*. To label 50 µg of protein with 400 pmol of dye:

1. Add 1 vol of protein sample equivalent to 50 µg to a microfuge tube.
2. Test the pH if this was not done earlier to confirm pH 8.5.
3. Add 1 µl of working dye solution (400 pmol) to the microfuge tube containing the protein sample.
4. Mix dye and protein sample thoroughly by pipetting and vortexing.
5. Centrifuge briefly in a microcentrifuge to collect the solution at the bottom of the tube.
6. Leave on ice for 30 min in the dark.
7. Add 1 µl of 10 mM lysine per 400 pmol of dye to stop the reaction.
8. Mix and spin briefly in a microcentrifuge.
9. Leave for 10 min on ice in the dark.

Labeling is now finished. The labeled samples can be processed immediately or stored for up to 3 months at –70 °C in the dark.

### 3.5. Preparation of Samples for IEF

The main difference between conventional 2-D electrophoresis and 2-D DIGE system is that the latter will enable up to three different protein samples to be run on a single 2-D gel. To achieve this, the differently labeled protein samples are combined before the first dimension run.

1. According to your experimental design, select a sample labeled with Cy3 and a sample labeled with Cy5 and combine the solutions in one microfuge tube (*see* **Table 4**). Use care when combining samples according to your experimental design.
2. Add an aliquot of the labeled pooled internal standard to the microfuge tube. The color of the combined solutions should be light purple. The internal standard must be divided equally across all pairings of samples. Use care to ensure you will have sufficient amount for the last tube!
3. Repeat **steps 1** and **2** for the remaining samples, according to your experimental design.
4. Add an equal volume of 2× sample buffer (7 M urea, 2 M thiourea, 2% CHAPS containing 130 mM DTT, and 2% ampholytes) to each of the combined samples and leave on ice for 15 min. The samples are now ready for IEF.

### *3.6. Isoelectric Focusing*

Once the sample(s) is prepared for IEF, the process for running the first dimension strip(s) is the same as it would be with traditional 2-DE *(27)*. Therefore, details regarding this process will not be outlined here.

### *3.7. Second Dimension*

The second dimension SDS–PAGE is also run the same as with traditional 2-DE, with one exception. Because of the use of fluorescent dyes, pre-cast or

**Table 4**
**Labeling Example**

|  | Protein Concentration (mg/ml) | Volume (μl) for 50 μg | Equalizing volume (lysis buffer) (μl) | Final volume for labeling | CyDye (μl) | Lysine (μl) |
|---|---|---|---|---|---|---|
| Cy2 | 10 | 5 | 5 | 10 | 1 | 1 |
| Cy3 | 5 | 10 | 0 | 10 | 1 | 1 |
| Cy5 | 7.5 | 6.7 | 3.5 | 10 | 1 | 1 |

Total volume (12 μl × 3 = 36 μl)
Add 2 × sample buffer = 36 μl
Final volume: 72 μl
A 24-cm IPGstrip needs a volume of 450 μl
Add 378 μl (450–72) of rehydration buffer to the combined sample
If the final concentrations vary, dilute the more concentrated samples with lysis buffer before labeling to ensure equivalent labeling in all reactions. After labeling, the combined sample is diluted with 2× sample buffer, and the volume increased with standard 2-D rehydration buffer to the volume required for the specified Immobiline™ DryStrip (IPGstrip).

self-cast gels with low fluorescence glass plates must be used to ensure high-quality results with minimal background. Gels cast with normal glass or plastic backing will have high background in one or more channels during scanning, compromising quantitation.

## 3.8. Image Acquisition

Images from each of the individual samples on each gel are obtained using either a laser-based scanner or a CCD-based imager. Fluorescent imaging ensures a wide dynamic range and excellent sensitivity. The imager or scanner should be capable of acquiring data in 16-bit TIFF format with 100 μm resolution for highest quality data, using narrow band filters for emission and excitation (or optimized laser wavelengths with appropriate filters on laser scanners) to avoid crosstalk between the dyes. The imager or scanner should also allow multi-channel image collection to provide seamless integration with DeCyder 2-D or ImageMaster Platinum software.

The basic steps for acquiring images will be described here, based on the use of Typhoon™ Variable Mode Imager, a multipurpose high-quality imager, or Ettan DIGE Imager, a dedicated fluorescent CCD scanner, both available from GE Healthcare. Both these imagers are capable of imaging gels still within their glass cassettes, which allows convenient handling, as well as immobilizing the gel for the required multiple scanning without worrying about gel movement, shrinkage, or swelling. For the Typhoon, the confocal optics allow the option of selecting a higher focal plane (+3 mm) to image within the glass cassette. Gels are scanned before fixing to ensure accurate quantitation.

1. Wipe outside of glass cassette to remove fingerprints, dust, and water spots. Use a lint-free wipe, such as Crew Wiper, to avoid introducing dust to the glass surfaces. Wear powder-free gloves when handling the glass cassettes.
2. Place one or more gel cassettes onto the scan area.
3. Set scanning criteria, including excitation and emission settings for each CyDye present, and resolution and scan area for each gel (*see* **Table 3**). For laser scanner, set focal plane for the laser and set each PMT at a level to avoid saturation. For CCD camera imagers, set the scan exposure time. Use settings for sensitivity, not speed, to achieve optimum results (*see* **Note 6**).
4. Perform pre-scan (if available at 500 or 1000 μm) to review that scanner settings are optimum. Make adjustments and repeat if necessary (*see* **Note 6**).
5. Perform final scan at 100 μm for large (25 × 20 cm) gels, 50 μm for smaller gels.
6. Review scan acquired to ensure no saturated peaks. Crop images as necessary to remove dye front or first dimension strip.
7. Repeat for remaining gels (*see* **Note 7**).

### *3.9. Image Analysis*

The images can now be analyzed using software capable of working with DIGE images. Although a full description of the use of such software is beyond the scope of this chapter, in brief, the software should be capable of accurate spot detection, normalization, and quantitation, followed by accurate spot matching and statistical analysis across multiple gels. The software should provide 2-D and 3-D viewing of the gel images, graphical views of the data from the sample groups, and the ability to select groups of spots to be picked for MS analysis based on spot coordinates. Two software platforms that are designed for this workflow are DeCyder 2-D and ImageMaster Platinum DIGE enabled, both from GE Healthcare. The focus here will be on the use of DeCyder 2-D, although both software packages use a spot detection algorithm designed to take full advantage of the multiplexing approach and the use of an internal standard on each gel.

This novel spot detection algorithm merges the images for each gel into one image, which is then used for spot detection. The images are re-separated, and the spot boundaries, or spot map, found during detection are then applied to each image, followed by calculation of each spot's area, volume and peak height, and normalization of each spot's volume. This approach ensures that for each gel, only one spot map is created for all three images, with identical spot boundaries across the images, providing an internal match between spots found on each image. The software also calculates the ratio between the internal standard in the Cy2 image and each of the individual samples labeled with Cy3 or Cy5 on the same gel. This provides normalization of the data across gels and eliminates gel-to-gel variability, resulting in more accurate quantitation between samples on different gels.

Each program then performs gel matching and statistical calculations. DeCyder 2-D uses the internal standard image on each gel to match across gels, reducing the number of individual matches required by two-thirds compared with traditional 2-D gels. Quantitation is calculated by using the internal ratio on each gel (C3/Cy2 or Cy5/Cy2) compared with the ratios on other gels. Using ratios with the internal standard ensures that the quantitation is accurate and virtually eliminates any contribution arising from gel-to-gel variation because the Cy2-labeled sample is included on each gel. Following gel matching, data can be analyzed through Average Ratio and *t*-test calculations for two sample comparisons, 1-way ANOVA for multiple samples, and 2-way ANOVA when multiple conditions (e.g., dose and time) are present in the experiment. DeCyder 2-D also has a module called Extended Data Analysis (EDA), with advanced statistical tools to provide detailed analysis of the results. These tools include principal component analysis, supervised and unsupervised clustering, and discriminant analysis to identify putative biomarkers of interest for classifying the samples. The ability to import information from gene ontology annotations related to biological function of identified proteins is also provided.

A key feature of DeCyder 2-D is the ability to compare not just two samples (e.g., control and treated) but an unlimited number of samples (e.g., time 0, time 1, time 2, …) within the same experiment. This increases the ability to design a complex experiment on the least number of gels and allow intra-sample comparisons between any groups within the same experiment. An example of an experiment containing multiple samples can be seen in **Fig. 3**. On the basis of the analysis, spots of interest can be assigned for spot picking and a pick list can be exported.

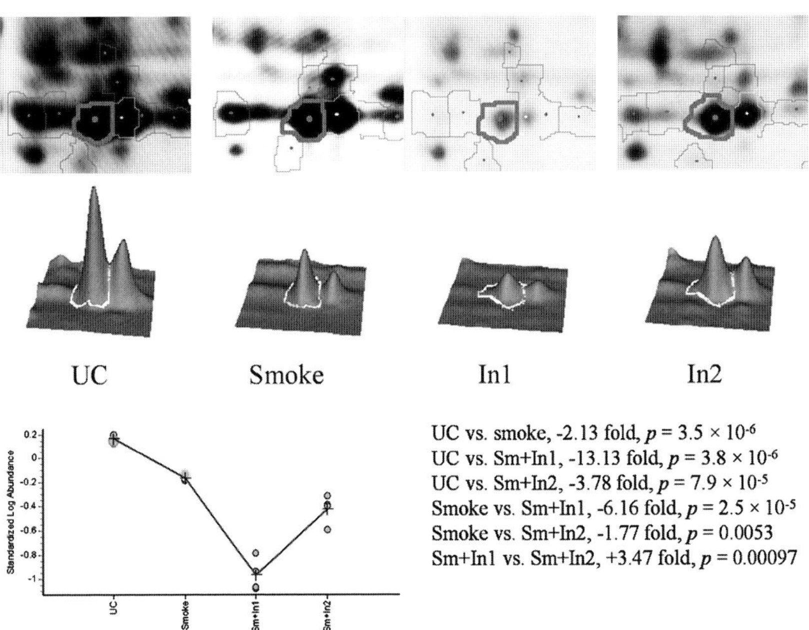



Fig. 3. Data from a 2-D difference gel electrophoresis (DIGE) experiment can be viewed in a number of ways in DeCyder 2-D, including 2-D spot maps, 3-D images visualizing specific spot intensities, and graphical depiction comparing the abundance of each matched spot across the samples in the experiment. This experiment compared lung tissue (universal control, UC) against tissue exposed to smoke (Smoke) and tissue exposed to smoke in the presence of an inhibitor (In1 or In2). All possible *t*-test and average ratio values can be calculated very easily as shown for this spot. Unpublished data from samples kindly provided by Dr. Koustubh Panda, Cleveland Clinic Foundation.

### *3.10. Post-Analysis Processing*

Spots of interest can be excised from a preparative gel (*see* **Note 3**), digested enzymatically, and prepared for MS analysis. This is generally done with either matrix-assisted laser desorption ionization (MALDI) instruments or electrospray liquid-chromatography (LC-MS) approaches. MS data can then be searched using any number of available search engines and protein databases to provide identifications (IDs) for the proteins from the picked spots. DeCyder 2-D allows the import of the IDs for annotation and, if using the EDA module, further investigation of the biological functions of those proteins shown to be differentially expressed among the samples in the experiment.

## 4. Notes

1. Standard 2-D gel equipment is used for 2-D DIGE, with the exception of low-fluorescence (LF) glass plates that are required for the second dimension. The use of LF glass plates ensures a low background during imaging so that the entire dynamic range can be used for quantitation. Keeping the gel within the glass plates also allows the ability to image the gel more easily than as a loose gel, and prevents shrinking, swelling, or movement of the gel while in the scanner tray during three consecutive scans for the three dyes.

2. The labeling is highly dependent on using anhydrous DMF for the dye solutions. Poor-quality DMF will result in reduced labeling and shelf life of the dyes, by causing premature hydrolysis of the $N$-hydroxysuccinimide ester on the minimal dyes. To avoid the formation of DMF amine-containing byproducts and the accumulation of water in the bottle, use a fresh bottle frequently, or add molecular sieves (Cat. No. M2635, Sigma Aldrich) to absorb the impurities and water. If the gel needs to be post-stained, the ethanol used for gel washing and fixing should be free of hexanes or other non-alcohol organic solvent impurities, which can fluoresce and increase background when scanning the gels for fluorescent post-stains. Ethanol containing small amounts of methanol or isopropanol is preferred.

3. 2-D DIGE allows the use of gels containing three samples, 50 µg each, for the analysis, resulting in 150 µg total protein on each gel. This is generally not sufficient material for downstream MS analysis after spot picking. To accommodate a higher protein load for MS analysis, one or two preparative-scale gels (containing unlabeled protein sample) are run alongside the analytical gels in the experiment or they can be run after image and statistical analysis is complete. Preparative gels can be prepared using a mixture of the samples, much the same way the internal standard is prepared. Because there is no quantitation performed on this gel, the goal is to simply have all possible proteins present for picking and MS analysis. A total of 500 µg or more is usually required to ensure ID of a majority of spots that will be picked, particularly for low-abundant proteins. Keep in mind that IEF

may require longer focusing time because of the higher amount of protein present compared with the analytical gels. Alternatively, the additional sample can be added as "cold" protein to the labeled mixture (300 μg or more added to the 150 μg of labeled mix) and included on one or more of the designated analytical gels [prepare one plate with Bind-Silane for picking purposes before casting this gel(s)]. For best spot picking, it is recommended to prepare preparative gels bound to one glass plate, treating the bottom plate with Bind-Silane before casting the gel. This ensures accurate spot picking by eliminating shrinking or swelling of the gel during the picking process. These preparative gels will have the same spot pattern, visualized using a fluorescent post-stain such as Deep Purple or SYPRO Ruby, and can be easily matched to the analytical gels. On the basis of the matching, spots of interest can then be placed in a pick list with the spot coordinates of the preparative gel.

4. The pH of the samples must be 8.5 for optimal labeling with minimal dyes. If a protein precipitation was performed with acidic conditions, adjustments are often needed to increase the pH, even after solubilization with the proper buffer. This can be done by using dilute NaOH (50–200 mM) and testing with pH paper. Incorrect pH of the sample will result in poor labeling.

5. The concentration of protein in all lysates should be determined using a protein assay (e.g., Ettan 2-D Quant Kit) that is compatible with typical 2-DE denaturing reagents present in the lysate solution. Protein concentration of 5–10 mg/ml in the lysate is ideal. However, samples containing 1 mg/ml protein have been successfully labeled using the protocol below. Use a protein precipitation method (e.g., Ettan 2-D Clean-up kit) to concentrate the sample(s) if necessary.

6. The scan settings required for DIGE gels must be optimized on both a laser scanner and a CCD imager. For the Typhoon laser scanner, the focal plane can be set to +3 mm to allow imaging between two low-fluorescence glass plates and the PMT must be set for each CyDye channel, typically between 500 and 600 V, the exact value determined by the pre-scan step. For the Ettan DIGE imager CCD-based imager, the exposure time must be optimized for each CyDye channel. The total scanning time for a 3-color DIGE large format (20 × 25 cm) gel is approximately 30 min on either platform. On the Typhoon, two large format gels can be scanned together in about 1 h.

7. Gels should be scanned at room temperature. Fluorescence is temperature dependent so all gels should be at the same temperature for equivalent scans. Gels can be stored at 4 °C if many gels are waiting to be scanned, even overnight, to minimize any spot diffusion before scanning, and scanning can be completed the next day if necessary. Allow cold gels to warm to room temperature before scanning.

## Acknowledgments

The authors thank Clayton Randall for his assistance preparing the figures, and Phil Beckett and Rita Marouga for critical review of the manuscript.

## References


1. Unlu, M., Morgan, M.E., and Minden, J.S. (1997) Difference gel electrophoresis: a single gel method for detecting changes in protein extracts. *Electrophoresis* **18**, 2071–77.
2. Kondo, T., Seike, M., Mori, Y., Fujii, K., Yamada, T., and Hirohashi, S. (2003) Application of sensitive fluorescent dyes in linkage of laser microdissection and two-dimensional gel electrophoresis as a cancer proteomic study tool. *Proteomics* **3**, 1758–66.
3. Zhou, G., Li, H., DeCamp, D., Chen, S., Shu, H., Gong, Y., Flaig, M., Gillespie, J. W., Hu, N., Taylor, P. R., Emmert-Buck, M. R., Liotta, L. A., Petricoin, E. F., III, and Zhao, Y. (2002) 2-D differential in-gel electrophoresis for the identification of esophageal scans cell cancer-specific protein markers. *Mol Cell Proteomics* **1**, 117–24.
4. Navet, R., Mathy, G., Douette, P., Dobson, R.L., Leprince, P., De Pauw, E., Sluse-Goffart, C., and Sluse, F.E. (2007) Mitoproteome plasticity of rat brown adipocytes in response to cold acclimation. *J Proteome Research* **6**, 25–33.
5. Freeman, W. M. and Hemby, S. E. (2004) Proteomics in protein expression profiling in neuroscience. *Neurochem Res* **29**, 1065–81.
6. Skynner, H. A., Rosahl, T. W., Knowles, M. R., Salim, K., Reid, L., Cothiliff, R., McAllister, G., and Guest, P. C. (2002) Alterations of stress related proteins in genetically altered mice revealed by two-dimensional in-gel electrophoresis analysis. *Proteomics* **2**, 1018–25.
7. Knowles, M. R., Cervino, S., Skynner, H. A., Hunt, S. P., de Felipe, C., Salim, K., Meneses-Lorente, G., McAllister, G., and Guest, P. C. (2003) Multiplex proteomic analysis by two-dimensional differential in-gel electrophoresis. *Proteomics* **3**, 1162–71.
8. Wilson, K. E., Marouga, R., Prime, J. E., D. Pashby, P., Orange, P. R., Crosier, S., Keith, A. B., Lathe, R., Mullins, J., Estibeiro, P., Bergling, H., Hawkins, E., and Morris, C. M. (2005) Comparative proteomic analysis using samples obtained with laser microdissection and saturation dye labeling. *Proteomics* **5**, 3851–58.
9. Somiari, R. I., Somiari, S., Russell, S., Craig, D., and Shriver, C. D. (2005) Proteomics of Breast Carcinoma. *J Chromatogr B Analyt Technol Biomed Life Sci* **815**, 215–25.
10. Somiari, R. I., Sullivan, A., Russell, S., Somiari, S., Hu, H., Jordan, R., George, A., Katenhusen, R., Buchowiecka, A., Arciero, C., Brzeski, H., Hooke, J., and Shriver, C. (2003) High-throughput proteomic analysis of human infiltrating ductal carcinoma of the breast. *Proteomics* **3**, 1863–73.
11. Friedman, D. B., Hill, S., Keller, J. W., Merchant, N. B., Levy, S. E., Coffey, R. J., and Caprioli, R., M. (2004) Proteome analysis of human colon cancer by two-dimensional difference gel electrophoresis and mass spectrometry. *Proteomics* **4**, 793–811.
12. Alfonso, P., Nunez, A., Madoz-Gurpide, J., Lonbardia, L., Sanchez, L., and Casal, J. I. (2005) Proteomic expression analysis of colorectal cancer by two-dimensional differential gel electrophoresis. *Proteomics* **5**, 2602–11.

13. Berendt, F. J., Frohlich, T., Schmidt, S. E., Reichenbach, H. D., Wolf, E., and Arnold, G. J. (2005) Holistic differential analysis of embryo-induced alterations in the proteome of bovine endometrium in the preattachment period. *Proteomics* **5**, 2551–60.

14. Greengauz-Roberts, O., Stöppler, H., Nomura, S., Yamaguchi, H., Goldenring, J. R., Podolsky, R. H., Lee, J. R., and Dynan, W.S. (2005) Saturation labeling with cysteine-reactive cyanine fluorescent dyes provides increased sensitivity for protein expression profiling of laser-microdissected clinical specimens. *Proteomics* **5**, 1746–57.

15. Kernec, F., Unlu, M., Labeikovsky, W., Minden, J. S, and Koretsky, A. P. (2001) Changes in the mitochondrial proteome from mouse hearts deficient in creatine kinase. *Physiol Genomics* **6**, 117–28.

16. Sakai, J., Ishikawa, H., Kojima, S., Satoh, H., Yamamoto, S., and Kanaoka, M. (2003) Proteomic analysis of rat heart in ischemia and ischemia-reperfusion using fluorescence two-dimensional difference gel electrophoresis. *Proteomics* **3**, 1318–24.

17. Chu, R., Lim, H., Brumfield, L., Liu, H., Herring, C., Ulintz, P., Reddy, J. K., and Davison, M. (2004) Protein profiling of mouse livers with peroxisome proliferator-activated receptor alpha activation. *Mol Cell Biol* **24**, 6288–97.

18. Ruepp, S. U., Tonge, R. P., Shaw, J., Wallis, N., and Pognan, F. (2002) Genomics and proteomics analysis of acetaminophen toxicity in mouse liver. *Toxicol Sci* **65**, 135–50.

19. Zhang, X., Guo, Y., Song, Y., Sun, W., Yu, C., Zhao, X., Wang, H., Jiang, H., Li, Y., Qian, X., Jiang, Y., and He, F. (2006) Proteomic analysis of individual variation in normal livers of human beings using difference gel electrophoresis. *Proteomics* **6**, 5260–8.

20. Prabakaran, S., Wengenroth, M., Lockstone, H.E., Lilley, K., Leweke, F.M., and Bahn, S. (2007) 2-D DIGE analysis of liver and red blood cells provides further evidence for oxidative stress in schizophrenia. *J Proteome Res* **6**, 141–9.

21. Seike, M., Kondo, T., Fujii, K., Okano, T., Yamada, T., Matsuno, Y., Gemma, A., Kudoh, S., and Hirohashi, S. (2005) Proteomic signatures for histological types of lung cancer. *Proteomics* **5**, 2939–48.

22. Seike, M., Kondo, T., Fujii, K., Yamada, T., Gemma, A., Kudoh, S., and Hirohashi, S. (2004) Proteomic signature of human cancer cells. *Proteomics* **4**, 2776–88.

23. Doran, P., Martin, G., Dowling, P., Jockusch, H., and Ohlendieck, K. (2006) Proteome analysis of the dystrophin-deficient MDX diaphragm reveals a drastic increase in the heat shock protein cvHSP. *Proteomics* **6**, 4610–21.

24. Van den Bergh, G., Clerens, S., Vandesande, F., and Arckens, L. (2003) Reversed-phase high-performance liquid chromatography prefractionation prior to two-dimensional difference gel electrophoresis and mass spectrometry identifies new differentially expressed proteins between striate cortex of kitten and adult cat. *Electrophoresis* **24**, 1471–81.

25. Van den Bergh, G., Clerens, S., Cnops, L., Vandesande, F., and Arckens, L. (2003) Fluorescent two-dimensional difference gel electrophoresis and mass spectrometry

identify age-related protein expression differences for the primary visual cortex of kitten and adult cat. *J Neurochemistry* **85**, 193–205.

26. Alban, A., David, S. O., Bjorkesten, L., Andersson, C., Sloge, E., Lewis, S. and Currie, I. (2003) A novel experimental design for comparative two-dimensional gel analysis: Two-dimensional difference gel electrophoresis incorporating a pooled internal standard. *Proteomics* **3**, 36–44.

27. Ettan DIGE System User Manual (2006) 18–1137–13 AB, GE Healthcare.

# 2

# Protein Carbohydrate Analysis

*Gel-Based Staining, Liquid Chromatography, Mass Spectrometry, and Microarray Screening*

Injae Shin, Alina D. Zamfir, and Bin Ye

## Summary

Carbohydrate modification of proteins and lipids serves to functionally "fine-tune" these molecules, which are involved in a wide variety of physiological and pathogenic processes through specific biological interactions. Therefore, errors in glycosylation have severe implications and have been associated with many common diseases such as cancer and diabetes. Carbohydrates have historically been desirable targets for drug invention. Recent advances in detection and analysis of carbohydrates have facilitated understanding of the roles of glycoproteins and carbohydrates in many cellular and signaling processes, paving the way for the exploitation of carbohydrates in disease diagnosis and drug discovery. Here, we introduce the most recently developed analytical methods, which have been routinely used in many research laboratories. Following the protein separation by one-dimensional (1D) and two-dimensional (2D) polyacrylamide gel electrophoresis (PAGE), gel-based staining methods provide the initial step to identify candidate glycoproteins so that further characterization can proceed. Combined liquid chromatography and mass spectrometry provide an invaluable technology for quantitative analysis of glycoproteins including glycopeptides and glycans and to identify the glycosylation site. The advanced technology of the carbohydrate-based microarray system incorporates high-throughput analysis to extend the scope of biomedical research on carbohydrate-mediated molecular interactions.

**Key Words:** Electrophoresis; gel-staining; glycoprotein; liquid chromatography; capillary electrophoresis; electrospray, mass spectrometry; carbohydrate microarrays; carbohydrate-protein interactions; diagnosis; enzymatic glycosylations; glycomics; high-throughput analysis; lectins.

From: *Methods in Molecular Biology, vol. 441: Tissue Proteomics: Pathways, Biomarkers, and Drug Discovery*
Edited by: B.C.-S. Liu © Humana Press, Totowa, NJ

## 1. Introduction

Advanced proteomic approaches including two-dimensional (2D)-gel separation, liquid chromatography-based protein fractionation, and mass spectrometry have great power to identify specific and novel gene targets and disease biomarkers. One-dimensional (1D) and 2D polyacrylamide gel electrophoresis (2D-PAGE) approaches are the most common methods for protein separation and characterization. Gel-based glycosylation analysis still provides an easy and quick method to view large scale of proteins (i.e., protein mixture) with glyco-modifications. In addition, from a brief literature survey, it appears that gel-based protein glycosylation analysis has several apparent advantages: *(1)* native glycoproteins, such as large size membrane glycoproteins can be separated by native and denatured gels, without further purification steps; *(2)* both 1D- and 2D-PAGE with subsequent blotting onto polyvinylidene difluoride (PVDF) and nitrocellulose membranes are compatible with downstream characterization such as antibody-based western blotting; *(3)* the gel-based analysis is compatible with other protein-staining methods such as silver staining and SYPRO Ruby staining and can be applied for further characterization using mass spectrometry technology after in-gel trypsin digestion. In particular, nanoelectrospray multiple stage MS (nanoESI-MS$^n$) in the positive as well as in the negative ion mode is capable of sequencing minute amounts of proteins and provides information on glycoprotein structure determinants *(1,2)*. The complete ESI-MS investigation of a glycoprotein is designed to determine *(1)* glycosylation sites; *(2)* the type of the monosaccharide moiety attached to the protein; *(3)* the sugar core type; *(4)* the type and size of the oligosaccharide chains attached to the core; *(5)* the branching patterns of carbohydrate chains; *(6)* the site of monosaccharide glycosidic linkages; *(7)* the covalent alteration of the sugar backbone chains by carbohydrate- and non-carbohydrate-type modifications; and *(8)* the non-covalent interactions. ESI-MS has the advantage of forming ions directly from solution. By online ESI high-performance liquid chromatography (HPLC) and CE, complex mixtures of glycopeptides or carbohydrates released from parent protein or even mixtures of intact glycoproteins can be separated and structurally characterized.

In an effort to develop a high-throughput tool to understand carbohydrate-mediated biological processes, carbohydrate microarrays have been prepared by immobilizing modified or unmodified carbohydrates on the proper solid surfaces *(3–18)*. These microarrays can be applied for the rapid analysis of carbohydrate-protein recognition events. The protein-binding patterns determined by these microarrays are consistent with those obtained from conventional solution-based assays. In addition, carbohydrate microarrays can be also useful for detecting pathogens for diagnosis of diseases because pathogens express specific carbohydrate-binding proteins on their surfaces.

These microarrays can be further utilized to determine quantitative binding affinities of lectins to carbohydrates and profile enzyme activities.

Overall, here, we provide three approaches, which are most frequently used in laboratories to analyze glycoproteins associated with biological regulation and disease biomarker/molecular medicine.

## 2. Materials

### 2.1. Materials for Gel-Based Staining

All chemicals used should be of analytical grade, the water should be pure, or de-ionization should be complete. The commercial staining methods used are Pro-Q Emerald 300 and Pro-Q Emerald 488 from Molecular Probes (Invitrogen, Carlsbad, CA, USA); GelCode glycoprotein staining kit from Pierce (Rockford, IL, USA); Dig Glycan detection kit from Roche Applied Science (Basel, Switzerland); ECL glycoprotein detection from Amersham-Pharmacia (GE Healthcare, Piscataway NJ, USA); and GlycoTrack detection kit from Glyko, Inc. (Novato, CA, USA). The materials listed below were obtained primarily from the glycoprotein gel and blot-staining kit, i.e., Pro-Q Emerald 300 (from Molecular Probes, Invitrogen).

1. Pro-Q Emerald 300 reagent (component A) at 50× in dimethyl formamide (DMF).
2. Pro-Q Emerald 300 staining buffer (component B), 250 ml.
3. Oxidizing reagent (component C): 2.5 g of periodic acid.
4. CandyCane glycoprotein molecular weight standards (component D).
5. DMF or dimethylsulfoxide (DMSO).
6. Methanol, glacial acetic acid, acetic acid.
7. Positive glycoprotein controls may include 1-acidic glycoprotein (40% carbohydrate), glucose oxidase (12% carbohydrate), and avidin (7% carbohydrate).
8. Enzyme conjugates such as horseradish peroxidase conjugate (HRP) and alkaline phosphatase conjugate (AP).
9. Plastic staining dishes.
10. Staining solution prepared with 10 ml Tris–HCl buffer, pH 9.5, and 200 µl NBT/X-phosphate solution.
11. Enzymes of *O*-glycanase™ (endo-α-*N*-acetylgalactosaminidase), recombinant *N*-glycanase™, PNGaseF, *N*-glycosidase F provided from Glyko, Inc.
12. Incubation buffer: 20 mM Sodium phosphate pH 7.5, with 50 mM ethylenediaminetetraacetic acid (EDTA) and 0.02% sodium azide.

### 2.2. Materials for LC/MS

All chemicals used should be of analytical grade and water should be deionized with high purity. HPLC-grade water is recommended for online LC/MS. The relevant chemical reagents and instruments are listed below:

                                                          

1. For LC/MS protocol: ammonium bicarbonate, hydrochloric acid, trypsin.
2. For CE/MS protocol: ammonium acetate, formic acid, ammonia.
3. Glycoprotein glycosylation sites may be assessed utilizing nano-LC interfaced to a Micromass QTOF Global (Waters, Milford, MA, USA).
4. CE Buffer (direct polarity): aqueous ammonium acetate/ammonia and 40% MeOH, pH 12.0.
5. CE Buffer (reverse polarity): 0.1 M formic acid in a methanol/water (6:4 v/v) solution adjusted with ammonia to pH 2.8.
6. The sample is separated using the Micromass CapLC (Waters) system.
7. A PACE 5000 series Capillary electrophoresis instrument (Beckman, Fullerton, CA, USA) may be used for CE/MS. For CE, externally polyimide-coated fused-silica capillaries of 75 μm inside diameter (ID) and 375 μm outside diameter (OD) are recommended. These capillaries may be obtained from BGB Analytic Vertrieb (Essen, Germany).
8. Sheathless CE/MS is performed on an orthogonal hybrid QTOF mass spectrometer (QTOF Micromass, Manchester, UK) equipped with a Z-spray ion source geometry. This mass spectrometer is interfaced with a PC running the Mass-Lynx N.T. software package to control the instrument and to record and process the MS data. The externally distal coated nanospray needles with 75 μm ID and 360 μm OD can be purchased from New Objective (Cambridge, MA, USA).

## *2.3. Materials for Carbohydrate Microarrays*

### *2.3.1. Labeling of Proteins by Fluorescent Dyes*

1. Fluorescein isothiocyanate (FITC) (Aldrich, St Louis, MO, USA or Molecular Probes, Eugene, OR, USA) is dissolved at 50 mg/ml in DMF or DMSO and stored at –20 °C. Because FITC is light-sensitive, keep the solution in the dark.
2. One tube of Cy3- or Cy5-*N*-hydroxysuccinimide (Cy3- or Cy5-NHS) (Amersham Biosciences, GE healthcare, Piscataway, NJ, USA) is dissolved in 10 μl DMSO and stored at –20 °C. Because Cy3 and Cy5 are light-sensitive, keep the solutions in the dark.
3. Unlabeled proteins are dissolved at 1–10 mg/ml (small proteins: 1–3 mg/ml, large proteins or antibodies: 1–10 mg/ml) in 0.1 M sodium phosphate buffer (pH 8.0) or 0.1 M sodium carbonate–sodium bicarbonate buffer (pH 8.0) and stored at 4 or –20 °C (*see* **Note 1**). Some fluorescent dye-labeled lectins are commercially available from Vector Laboratories, Burlingame, CA, USA or Sigma, St Louis, MO, USA.
4. PD-10 desalting column (Amersham Biosciences, GE healthcare, Piscataway, NJ, USA).

### *2.3.2. Fabrication of Carbohydrate Microarrays*

1. Buffers for dissolving carbohydrate probes: phosphate-buffered saline (PBS, pH 6.8) or 0.1 M sodium phosphate buffer (pH 5.0) is prepared and glycerol is added to the buffer (v/v, 40–50%) (*see* **Note 2**). The pH value of the glycerol-containing buffer is re-adjusted to the desired value (*see* **Note 3**).

2. Maleimide- and hydrazide-conjugated carbohydrates are not commercially available and thus should be chemically synthesized *(3–5,13)*. The maleimide- or hydrazide-conjugated carbohydrate probes are dissolved at 0.1–5.0 mM in PBS (pH 6.8) or 0.1 M sodium phosphate buffer (pH 5.0) containing 40–50% glycerol, respectively, and then stored at –70 °C. If solubility of the probes is poor, dissolve the probes in a minimum amount of DMF, and dilute with the proper buffer containing 40–50% glycerol.
3. Amine- or thiol-coated glass slides (*see* **Note 4**). Amine glass slides can be purchased from commercial suppliers such as TeleChem International, Inc., Sunnyvale, CA, USA or Schott Nexterion, Mainz, Germany.
4. 1% *N*-Ethylmaleimide (Sigma, St Louis, MO, USA) in water to quench the unreacted thiol groups after immobilization of carbohydrate probes on the thiol-coated glass slide.
5. 3% Poly(ethylene glycol)diglycidyl ether (Aldrich) in 10 mM $NaHCO_3$ (pH 8.3) to prepare epoxide-coated glass slides from amine slides.
6. PBS (pH 7.4) containing 0.1% Tween 20 (Aldrich).

## 2.3.3. Applications of Carbohydrate Microarrays

1. PBS (pH 7.4) containing 0.1% Tween 20 and 1% bovine serum albumin (BSA). BSA should be added to the buffer before use.
2. Enzymatic galactosylation: a solution of β-1,4-galactosyltransferase (23 mU) (Calbiochem), 10 mM $MnCl_2$, and 0.1 mM UDP-Gal in 50 mM HEPES buffer, pH 7.5.
3. Enzymatic sialylation: a solution of α-2,3-sialyltransferase (1 mU) (Calbiochem, San Diego, CA, USA), 5 mM $MnCl_2$, AP (20 µU), and 0.1 mM CMP-NeuNAc in 100 mM HEPES buffer, pH 7.0.
4. Enzymatic fucosylation: a solution of α-1,3-fucosyltransferase (1 mU) (Calbiochem), 15 mM $MnCl_2$, AP (20 µU) and 0.1 mM GDP-Fuc in 50 mM MES buffer, pH 6.0.
5. Mouse anti-sialyl $Le^x$ antibody (5–10 µg/ml) (Calbiochem) in 10 mM sodium phosphate (pH 7.2) containing 500 mM NaCl and 0.02% Tween 20. Tween 20 should be added to the solution before use.
6. 10 mM sodium phosphate (pH 7.2) containing 500 mM NaCl and 0.02% Tween 20 for washing unbound anti-sialyl $Le^x$ antibody.
7. Goat anti-antibody (10 µg/ml) (Calbiochem) in 10 mM sodium phosphate (pH 7.2) containing 500 mM NaCl and 0.02% Tween 20. Tween 20 should be added to the solution before use.

# 3. Methods

## 3.1. Methods of Gel-Based Protein Carbohydrate Staining

### 3.1.1. Pro-Q Emerald 300 Dye Staining

This green fluorescent Pro-Q Emerald 300 glycoprotein-staining method can be combined with SYPRO Ruby for total protein staining, with sensitivity up to

1 ng/band. Overall, the specificity of glycoprotein detection using this method depends greatly on adequate fixation and washing to remove sodium dodecyl sulfate (SDS) from the proteins and washing after the oxidation reaction.

1. Separate proteins by standard SDS–polyacrylamide gel electrophoresis (PAGE) on either 1D or 2D gels. Typically, the sample is diluted to about 10–100 µg/ml of sample buffer. Non-denatured native gel-separated proteins may be applied. Large 2D gels require proportionally larger volume and longer fixation and staining times.
2. After electrophoresis, fix the gel by immersing it in 50–100 ml of fixing solution and incubate at room temperature with gentle agitation for 45 min.
3. Wash the gel or blot by incubation with approximately 100 ml of wash solution with gentle agitation for 10 min. Repeat this step once.
4. Oxidize the carbohydrates. Incubate the gel or blot in 25 ml oxidizing solution with gentle agitation for 30 min.
5. Wash the gel or blot in 50 ml of wash solution with gentle agitation for 10–20 min and repeat this step twice more (three times more for large gels).
6. Prepare fresh Pro-Q Emerald 300 staining solution by diluting the Pro-Q Emerald 300 reagent (component A) 50-fold into Pro-Q Emerald 300 dilution buffer (component B).
7. Stain the gel or blot with incubation in 25 ml Pro-Q Emerald 300 staining solution in the dark for 90–120 min. Overnight incubation is not recommended.
8. Wash the gel with 50 ml of washing solution at room temperature for 15 min. Repeat this wash step once.
9. Visualize the stain using a standard UV transilluminator (*see* **Notes 5–6**).

### 3.1.2. GelCode Glycoprotein Staining

A similar staining method using Pierce's GelCode Glycoprotein staining kit is described below:

1. After electrophoresis, fix gel by completely immersing it in 100 ml of 50% methanol for 30 min.
2. Wash gel by gently agitating with 100 ml of 3% acetic acid for 10 min, repeat this step (This gel can be left in water overnight at 4 °C).
3. Transfer gel to 25 ml oxidizing solution and gently agitate for 15 min.
4. Wash gel by gentle agitation with 100 ml 3% acetic acid for 5 min, and repeat this step twice.
5. Transfer gel to 25 ml of GelCode glycoprotein-staining reagent and incubate for 15 min (if crystals form in the reagent, remove them by centrifugation; do not use heat to dissolve crystals).
6. Transfer gel to 25 ml of reducing solution and agitate for 5 min.
7. Wash gel extensively with 3% acetic acid and then with ultra-pure water; store in 3% acidic acid. Glycoproteins appear as magenta bands.

### 3.1.3. DIG Glycan Detection

The DIG Glycan detection method, provided by Roche Applied Science, is applied to transferred blot on nitrocellulose and PVDF membrane. Similar to the GlycoTrack (Glyko, Inc.) and Immun-Blot Glycoprotein (Bio-Rad, Richmond, Hercules, CA) detection kits, these staining methods involve protein conjugation and antibody for the detection of glycoproteins and may require a blocking step and a longer time for staining.

1. Oxidation and labeling with digoxigenin:

    A. dissolve 0.1–10 µg protein in 20 µl sodium acetate buffer, pH 5.5.
    B. Add 2 µl sodium metaperiodate solution and mix. Incubate for 20 min in dark at 15–25 °C.
    C. Destroy excess periodate by adding 10 µl sodium disulfite and leave at 15–25 °C for 5 min.
    D. Add 5 µl DIG-3-O-succinyl-ε-aminocaproic acid hydrazide, mix, and incubate at 15–25 °C for 1 h.
    E. Add SDS sample buffer, heat the mixture at 100 °C for 2 min, and apply to SDS–PAGE.

2. After electrophoresis is complete and the gel transferred onto PVDF membrane, immerse the membrane in the blocking solution for 30 min. If necessary, it can be incubated overnight at 4 °C.
3. Wash with 50 ml TBS buffer for 10 min three times.
4. Incubate with anti-digoxigenin-AP, add 10 µl conjugate to 10 ml TBS, and incubate for 1 h.
5. Repeat **step 3**, washing three times with 50 ml TBS buffer.
6. Stain with the fresh prepared staining solution.
7. Stop reaction when gray and black spots appear by rinsing with water (this reaction can usually be completed within a few minutes to overnight, depending upon the amount of glycoprotein).
8. Visualize the staining gel or membrane and document the image with CCD camera or other available software (*see* **Note 7**).

### 3.1.4. N- and O-Glycanase Coupled with Gel Staining

To define the specific type of glyco-modification of glycoproteins in biological samples, enzymatic treatment with *N*- and *O*-glycanases is very useful in parallel with the above-described glycoprotein-staining method. *N*-glycanase produces site-specific cleavage between asparagine and the proximal *N*-acetyl glucosamine of most oligomannoses, hybrid and complex type *N*-glycans. *O*-glycanase is highly specific, cleaving Galβ1, 3GalNAc from serine or threonine residues of glycoproteins. Comparing the protein sample with and without *N*- and *O*-glycanase pretreatment followed by protein separation, and glyco-staining, the specific types of glyco-modification differentiate the specimens with different disease conditions *(19,20)*.

1. Prepare 10–100 μg total protein sample in 45 μl of 1× incubation buffer. Add 2.5 μl of SDS/β-mercaptoethanol to final concentration 0.1% SDS and 50 mM β-mercaptoethanol.
2. Denature glycoprotein by heating at 100 °C for 5 min and allow the mixture to cool.
3. Add 2.5 μl NP-40 to final reaction concentration 0.75% NP-40.
4. Add 2 μl *N*-glycanase to the reaction mixture and incubate for 2 h to overnight at 37 °C.
5. Apply similar procedures for *O*-glycanase but omitting SDS and heat treatment.
6. Protein samples with and without enzyme treatment are applied for SDS-gel separation and glycoprotein staining as described above.

## 3.2. Methods for Glycoprotein Analysis Using Liquid Chromatography and Capillary Electrophoresis/Electrospray Mass Spectrometry

### 3.2.1. Determination of Glycosylation Sites by LC/ESI MS and MS/MS

For the determination of the glycosylation sites and assessment of the heterogeneity of the glycopeptides resulting from protein digestion, protein samples ought to be screened and sequenced under appropriate ESI MS conditions in combination with an efficient separation technique before MS *(21)*. The digested mixture will have the following characteristics: *(1)* site heterogeneity and/or adduct ion formation; *(2)* suppression of the glycopeptide signals in the presence of non-glycosylated peptides; *(3)* glycan portion heterogeneity; *(4)* multiple charging formation resulting in a global glycopeptide signal allocation/distribution onto several MS peaks with diminished intensity of single peak. For these reasons, it is necessary to use microcolumn liquid chromatography (μLC) because of its *(1)* low sample requirement; *(2)* choice of selectivity modes; *(3)* high separation efficiency; *(4)* ease of implementation; *(5)* commercially available coupling interfaces, fully functional or readily adaptable to any ESI MS ion source configuration.

#### 3.2.1.1. TRYPTIC DIGESTION OF GLYCOPROTEINS

1. Resuspend 1.0 mg of glycoprotein in 100 mM ammonium bicarbonate, pH 8.6, and thermally denature the glycoprotein.
2. Prepare a trypsin stock solution by dissolving 20 mg of the enzyme in 1 mM hydrochloric acid aqueous solution to a 1-mg/ml final concentration.
3. Dilute an aliquot of this solution 50-fold.
4. Add 1 ml of the solution obtained in **step 3** to the denatured glycoprotein and incubate overnight at 37 °C.

### 3.2.1.2. GLYCOSYLATION SITE DETERMINATION BY μLC/ESI QTOF MS AND MS/MS

1. Introduce 0.1 mg of the typically digested sample into LC using the Micromass CapLC system (*see* **Note 8**).
2. Configure the stream select module with a trapping C18 column where the sample will be loaded and desalted.
3. Use a 60-min gradient (0–40%) B (97% acetonitrile and 0.1% formic acid) to initiate the elution and separate the sample.
4. Set a 250-nl/min flow by using precolumn splitter.
5. Within parent ion discovery (PID) experiment, switch the voltage on the gas collision cell between 35 and 8 V every second.
6. Upon detection of the carbohydrate-characteristic oxonium ions, switch the QTOF instrument to MS/MS mode (DDA) and select the most intensely charged ion (triply or quadruply charged) for fragmentation.
7. Perform MS/MS DDA for 6 s at 1-s scan rate.
8. Repeat until the eight most intense precursor ions during a single scan are selected for MS/MS experiments.
9. During the MS/MS, apply a collision energy ramp from 20–40 V.

### 3.2.2. Glycopeptide Analysis by CE/ESI-MS

The release of *O*- and *N*-glycopeptides from the parent protein can also be analyzed by CE/ESI MS, which has the advantage of high resolution, low sample and electrolyte consumption (nanoliters to microliters per analysis), and high resolving power and separation efficiency *(22–24)*. However, attention must be paid to choosing the appropriate polarity and the corresponding buffer system and instrumental parameters.

### 3.2.2.1. DIRECT POLARITY

1. Construct the sheathless interface by attaching the CE separation capillary to the commercial distal coated nanospray needle through a teflon joint. The length of the CE capillary is to be set to 100–130 cm.
2. Incorporate the resulting column into a stainless steel-clenching device, designed and constructed to allow the application of the ESI voltage to the needle (*see* **Fig. 1**, **Note 9**).
3. Remove the conventional ESI source and screw the assembly onto the QTOF high-voltage plate.
4. Before sample application, condition the CE column by rinsing with 19 M aqueous ammonium hydroxide for 20 min followed by flushing with the CE buffer (direct polarity) for 20 min.
5. Dissolve the sample in the CE buffer (direct polarity) to a concentration of 0.5–0.75 mg/ml and hydrodynamically inject it into the CE column (*see* **Note 8**). Use an injection time of 8 s, which will result in an injected volume of 25 nl.

6. Separate the mixture and subject it to online ESI MS detection under 30 kV CE voltage applied on the anode, –1.1 kV (negative ion mode) on the cathode (ESI tip), and 40V on the MS counterelectrode (*see* **Note 10**).
7. Use desolvation gas (nitrogen) at 50 l/h for all experiments and generate the extracted ion chromatograms (XIC) within a mass window of 0.3 kDa for the most abundant glycopeptide ion in the total ion chromatogram (TIC) MS.
8. Program the QTOF instrument to record the data at a scan speed of 2.1 scans/s.
9. After generating the spectra, recalibrate them using NaI as the calibrant.

### 3.2.2.2. REVERSE POLARITY

1. Couple the CE and MS instruments through the sheathless interface by attaching the CE separation capillary to the commercial distal coated nanospray needle through a teflon joint (*see* **Note 11**).
2. Incorporate the resulting column into the stainless-steel-clenching device (*see* **Fig. 1**, **Note 9**).
3. Set the length of the CE capillary to 100–130 cm.
4. Dissolve the sample to the concentration of 0.5–0.75 mg/ml in CE buffer (reverse polarity); hydrodynamically inject it into the CE column by applying the injection pressure for 4–10 s (*see* **Note 8**).
5. Carry out the CE separation in reverse polarity by application of –25 kV separation voltage on the inlet (cathode) electrode (*see* **Note 12**).
6. Apply the ESI voltage on the ESI tip (anode) after the application of the CE separation voltage and initiate the electrospray at values of 800–1000 V in the negative ion mode (*see* **Note 10**).
7. Carry out **steps 8**, **9**, and **10** as described for direct polarity.

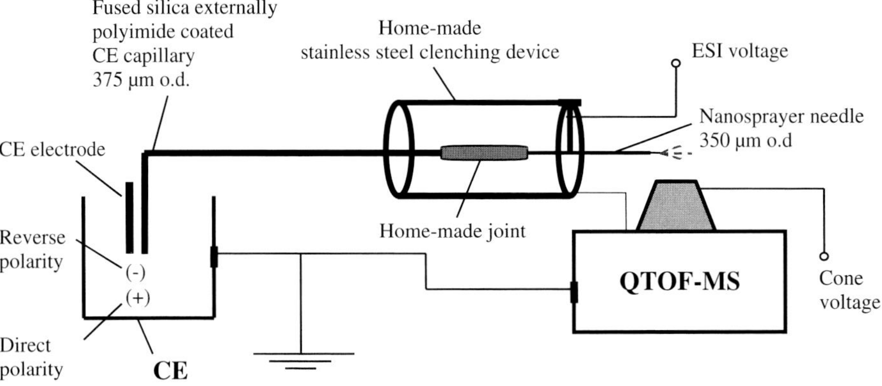


Fig. 1. Scheme of the sheathless CE/ESI MS interface for direct and reverse polarity.

## 3.3. Methods for Carbohydrate Microarrays

Carbohydrate microarrays are fabricated by printing very small quantities (1–20 nl) of carbohydrate probes on the proper solid surfaces with a robotic printing microarrayer. Carbohydrate probes should be dissolved in the proper buffer containing 40–50% glycerol to prevent undesired evaporation of nanodroplets during spotting and immobilization steps. A variety of fabrication methods have been previously reported *(3–18)*. Among them, we describe the fabrication procedure to immobilize maleimide- and hydrazide-conjugated carbohydrates on the thiol- and epoxide-coated glass slides, respectively *(3–5)* (*see* **Fig. 2**).

Fabricated microarrays have been used for biological research and biomedical applications. For these studies, the carbohydrate microarrays are incubated with fluorescent dye-labeled proteins or cells, and bound proteins or cells are visualized or quantitated by using a fluorescence scanner. A commonly occurring problem in the detection of bound proteins on carbohydrate microarrays is a high-background fluorescence signal because of non-specific interactions of fluorescent proteins with modified surfaces. This problem can be solved by treating the fabricated carbohydrate microarrays with 1% BSA before incubation of the microarrays with labeled proteins or cells. Alternatively, hydrophilic surfaces coated by poly(ethylene glycol) considerably reduce the non-specific adsorption of proteins on the surfaces *(11)*.

### 3.3.1. Fabrication of Carbohydrate Microarrays

3.3.1.1. Immobilization of Maleimide-Conjugated Carbohydrates on Thiol-coated Glass Slides

1. Dissolve maleimide-conjugated carbohydrate probes (0.1–5.0 mM) in PBS (pH 6.8) containing 40–50% glycerol. In the case of poor solubility of the probes, dissolve the probes in a minimum amount of DMF and dilute with the buffer containing 40–50% glycerol.

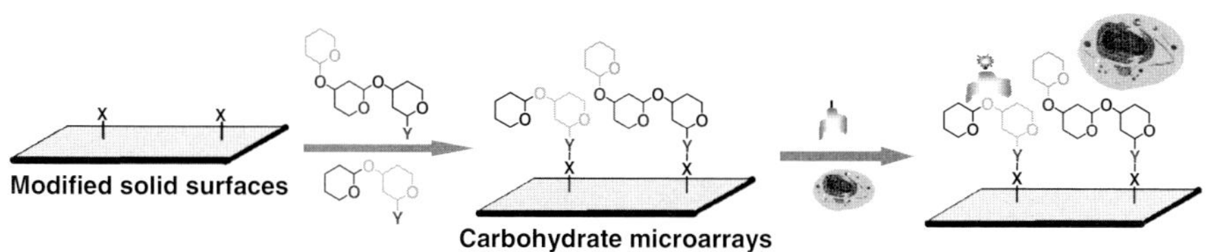


Fig. 2. Fabrication of carbohydrate microarrays and their applications for detection of proteins or cells. (a) X = -SH, Y = tethered maleimide *(3,4)* and (b) X = epoxide, Y = tethered hydrazide *(5)*.

2. Print 1–2 nl of carbohydrate solutions from a 384-microwell plate on a thiol-coated glass slide using a robotic printing microarrayer.
3. Place the printed slide in a humid chamber (50–70% humidity) at room temperature for 5–10 h. The low humidity causes evaporation of nanodroplets.
4. Briefly wash the printed slide with water and immerse in water containing 1% *N*-ethylmaleimide with gentle shaking for 15 min at room temperature to remove unreacted thiol groups (*see* **Note 13**).
5. Wash the slide with PBS (pH 7.4) containing 0.1% Tween 20 with gentle shaking for 15 min three times and rinse with water.
6. Dry the slide by purging with Ar gas and store in a desiccator until use.

### 3.3.1.2. IMMOBILIZATION OF HYDRAZIDE-CONJUGATED CARBOHYDRATES ON EPOXIDE-COATED GLASS SLIDES

1. Prepare epoxy-coated glass slides: place amine-coated glass slides in a staining jar containing 3% poly(ethylene glycol)diglycidyl ether in 10 mM NaHCO$_3$ (pH 8.3) and gently shake the jar for 30 min at room temperature. After thorough washing with water, dry the slides by purging with Ar gas and store in a desiccator until use.
2. Dissolve hydrazide-conjugated carbohydrate probes (0.1–5.0 mM) in 0.1 M sodium phosphate buffer (pH 5.0) containing 40–50% glycerol. In the case of poor solubility of the probes, dissolve the probes in a minimum amount of DMF and dilute with the buffer containing 40–50% glycerol.
3. Print 1–2 nl of carbohydrate solutions from a 384-microwell plate on an epoxy-coated glass slide using a robotic printing microarrayer.
4. Place the printed slide in a humid chamber (50–70% humidity) for 3–4 h at room temperature.
5. Briefly wash the printed slide with water and immerse in 10 mM PBS (pH 7.4) containing 0.1% Tween 20 with gentle shaking at room temperature for 15 min three times (*see* **Note 14**).
6. Rinse the slide with water.
7. Dry the slide by purging with Ar gas and then store in a desiccator until use.

### 3.3.2. Labeling of Proteins or Bacteria by Fluorescent Dyes

### 3.3.2.1. LABELING OF PROTEINS BY FLUORESCEIN

1. Fluorescein-labeled proteins are obtained by reacting proteins with FITC (*see* **Fig. 3**). Dissolve the protein (small proteins: 0.1–0.3 mg, large proteins or antibodies: 0.1–1 mg)in 100 µl of 0.1 M sodium phosphate buffer (pH 8.0). If protein activity is decreased by phosphate ions, the labeling reaction can be carried out in 0.1 M bicarbonate, HEPES, or borate buffers at the same pH. However, do not use buffers containing free amino groups such as Tris or glycine; these buffers will react with FITC.
2. Add 2 µl of a solution of FITC (50 mg/ml) in DMF or DMSO to the protein solution.

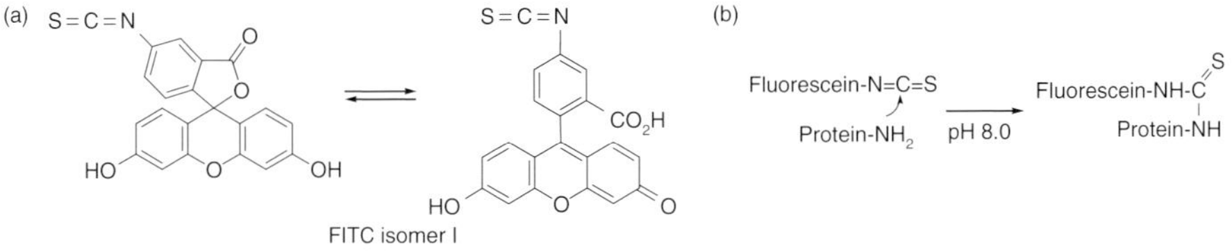

Fig. 3. (a) Structure of FITC. (b) Labeling of proteins by FITC. Amino groups of lysine residues of proteins react with FITC to generate the stable thiourea linkage.

3. Mix thoroughly and incubate the reaction mixture for 0.5 h at room temperature (*see* **Note 15**). Be careful not to foam the protein solution.

4. Adjust the volume of the reaction mixture to 2.5 ml with protein storage buffer and then load on the PD-10 desalting column that is pre-equilibrated with approximately 25 ml protein storage buffer (*see* **Note 16**). Discard the flow-through.

5. Elute with 3.5 ml protein storage buffer and collect the flow-through. Two colored bands are usually observed; the faster-moving band contains the labeled protein and the slower band contains free dye.

6. Store the labeled proteins at 4 or −20 °C. The excitation and emission wavelengths of fluorescein are 492 and 520 nm, respectively. The fluorescence properties of a labeled protein and free dye are very similar.

### 3.3.2.2. LABELING OF PROTEINS BY CY3 OR CY5

1. Cy3- or Cy5-labeled proteins are obtained by reacting proteins with Cy3-NHS or Cy5-NHS, respectively. Dissolve the protein (small proteins: 0.1–0.3 mg, large proteins or antibodies: 0.1–1 mg) in 100 μl of 0.1 M sodium carbonate–sodium bicarbonate buffer (pH 8.0). Do not use buffers containing free amino groups such as Tris or glycine. These buffers will react with Cy3-NHS and Cy5-NHS.

2. Add 1 μl of a solution of Cy3-NHS or Cy5-NHS in DMSO to the protein solution.

3. Mix thoroughly and incubate the reaction mixture for 30 min at room temperature. Be careful not to foam the protein solution.

4. Adjust the volume of the reaction mixture to 2.5 ml with protein storage buffer and then load on the PD-10-desalting column that is pre-equilibrated with approximately 25 ml protein storage buffer (*see* **Note 16**). Discard the flow-through.

5. Elute with 3.5 ml protein storage buffer and collect the flow-through. Two colored bands are usually observed; the faster-moving band contains the labeled protein and the slower band contains free dye.

6. Store the labeled proteins at 4 or −20 °C. The excitation and emission wavelengths of Cy3 are 550 and 570 nm, respectively, and the excitation and emission wavelengths of Cy5 are 649 and 670 nm, respectively. The fluorescence properties of a labeled protein and free dye are very similar.

### 3.3.2.3. Labeling of Bacteria by Fluorescent Dyes

1. Grow bacterial cells in Luria-Bertani medium at 37 °C and harvest the cells by centrifugation.
2. Suspend the collected cells with PBS (pH 7.4) and then centrifuge again. Repeat this step twice.
3. Re-suspend the cells with PBS (pH 7.4) and add 10 µl propidium iodide (1 mg/ml) dissolved in water or 1 µl SYTO 83 (5 mM) dissolved in DMSO to the suspended cells (100 µl) (*see* **Note 17**).
4. After incubation for 0.5–1 h at room temperature with gentle shaking, use the mixture without any manipulation for detection of the cells with carbohydrate microarrays (*see* **Note 18**). The excitation and emission wavelengths of propidium iodide bound to nucleic acid are 530 and 620 nm, respectively, and the excitation and emission wavelengths of SYTO 83 bound to nucleic acid are 543 and 559 nm, respectively.

## 3.3.3. Applications of Carbohydrate Microarrays

### 3.3.3.1. Analysis of Lectin–carbohydrate Interactions

1. Before using the carbohydrate microarray, immerse a slide into PBS (pH 7.4) containing 0.1% Tween 20 and 1% BSA with gentle shaking for 30 min at room temperature (*see* **Note 19**).
2. Wash the slide with PBS (pH 7.4) containing 0.1% Tween 20 with gentle shaking for 15 min three times and rinse with water.
3. Dry the slide by purging with Ar gas.
4. Incubate the carbohydrate microarray with dye-labeled proteins (1–10 µg/ml) in PBS (pH 7.4) containing 0.1% Tween 20 for 0.5–1 h at room temperature (*see* **Note 20**). For *ConA* binding, add $MnCl_2$ and $CaCl_2$ at final concentrations of 0.1–1 mM.
5. Remove the unbound proteins by washing the slide with the same buffer with gentle shaking for 15 min three times and rinse with water.
6. Dry the slide by purging with Ar gas and detect bound proteins using a fluorescence scanner (*see* **Fig. 4**).

### 3.3.3.2. Detection of Pathogens

1. Before using the carbohydrate microarray, immerse a slide into PBS (pH 7.4) containing 0.1% Tween 20 and 1% BSA with gentle shaking for 30 min at room temperature.
2. Wash the slide with PBS (pH 7.4) containing 0.1% Tween 20 with gentle shaking for 15 min three times and rinse with water.
3. Dry the slide by purging with Ar gas.
4. Pour the stained cells prepared by the procedure in **step 4** of **Subheading 3.3.2.3** onto the slide *(6,16,17)* (*see* **Note 18**).
5. After incubation at room temperature for 30 min, remove unbound cells by gentle washing with PBS (pH 7.4) and rinse with water.
6. Dry the slide by purging with Ar gas and detect bound cells using light microscopy or a fluorescence scanner.

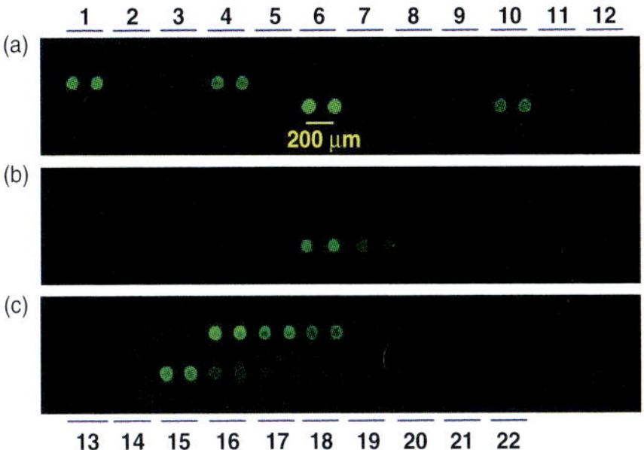

Fig. 4. Carbohydrate microarrays containing 22 carbohydrate probes after incubation with (a) FITC-labeled *Con A*, (b) FITC-labeled *N. pseudonarcissus* and (c) FITC-labeled *wheat germ agglutinin*.

### 3.3.3.3. QUANTITATIVE ANALYSIS OF LECTIN BINDING TO IMMOBILIZED CARBOHYDRATES BY DETERMINING $IC_{50}$ VALUES

1. Prepare the carbohydrate microarray immobilized by carbohydrate probes whose binding affinities are determined, according to the procedure described in **Subheading 3.3.1.1**.
2. Immerse the slide in PBS (pH 7.4) containing 0.1% Tween 20 and 1% BSA with gentle shaking for 30 min at room temperature.
3. Wash the slide with PBS (pH 7.4) containing 0.1% Tween 20 with gentle shaking for 15 min three times and rinse with water.
4. Dry the slide by purging with Ar gas.
5. Incubate the carbohydrate microarrays with a series of mixtures of a dye-labeled protein and a soluble inhibitor in PBS (pH 7.4) containing 0.1% Tween 20 for 1 h at room temperature.
6. Wash the slide with PBS (pH 7.4) containing 0.1% Tween 20 with gentle shaking for 15 min three times and rinse with water.
7. Dry the slide by purging with Ar gas and measure fluorescence intensities using a fluorescence scanner. A sigmoid curve is obtained; $x$-axis: log (concentration of an inhibitor), $y$-axis: fluorescence intensity.
8. Determine $IC_{50}$ values of inhibitors from the obtained graph *(4,11)*.

### 3.3.3.4. ENZYMATIC GLYCOSYLATIONS TO PREPARE SIALYL $Le^x$ FROM GLCNAC-IMMOBILIZED GLASS SLIDES.

1. Prepare the carbohydrate microarray containing GlcNAc according to the procedure described in **Subheading 3.3.1.1** (*see* **Fig. 5**).
2. For galactosylation of GlcNAc, incubate the slide with 15 µl of a solution of β-1,4-galactosyltransferase (GalT, 23 mU), $MnCl_2$ (10 mM), and UDP-Gal (0.1 mM) in HEPES buffer (50 mM, pH 7.5) in a humid chamber (50–70%) for 15 h at 37 °C.
3. Wash the slide with PBS (pH 7.4) containing 0.1% Tween 20 with gentle shaking for 15 min three times and rinse with water.
4. Dry the slide by purging with Ar gas.
5. For sialylation of LacNAc, incubate the slide with 15 µl of a solution of α-2,3-sialyltransferase (SialT, 1 mU), $MnCl_2$ (5 mM), AP (20 µU), and CMP-NeuNAc (0.1 mM) in HEPES buffer (100 mM, pH 7.0) in a humid chamber (50–70%) for 15 h at 37 °C.
6. Wash the slide with PBS (pH 7.4) containing 0.1% Tween 20 with gentle shaking for 15 min three times and rinse with water.
7. Dry the slide by purging with Ar gas.
8. For fucosylation of NeuNAcα2,6LacNAc, incubate the slide with 15 µl of a solution of α-1,3-fucosyltransferase (FucT, 1 mU), $MnCl_2$ (15 mM), AP (20 µU), and GDP-Fuc (0.1 mM) in MES buffer (50 mM, pH 6.0) in a humid chamber (50–70%) for 15 h at 37 °C.
9. Wash the slide with PBS (pH 7.4) containing 0.1% Tween 20 with gentle shaking for 15 min three times and rinse with water.

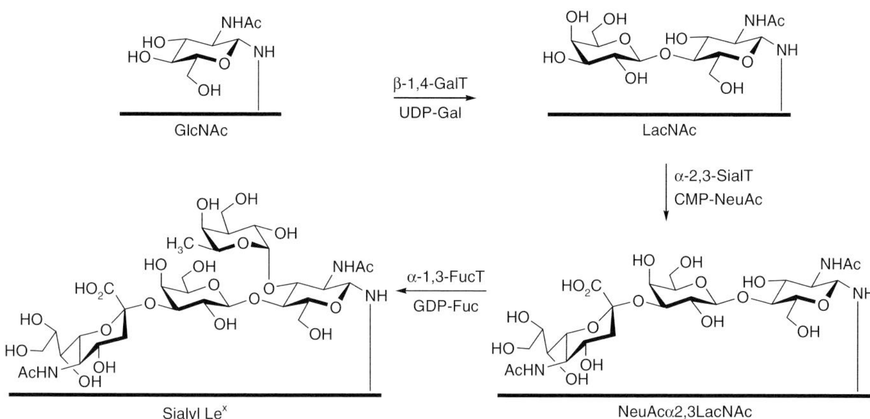


Fig. 5. Enzymatic synthesis of sialyl Le$^x$ from GlcNAc by three glycosyltransferases.

10. Dry the slide by purging with Ar gas.
11. To determine successful glycosylation, incubate the slide with PBS (pH 7.4) containing 0.1% Tween 20 and 1% BSA for 30 min at room temperature and then treat with mouse anti-sialyl Le$^x$ antibody (5–10 µg/ml) in 10 mM sodium phosphate (pH 7.2) containing 500 mM NaCl and 0.02% Tween 20 for 1 h at room temperature.
12. Wash the slide with 10 mM sodium phosphate (pH 7.2) containing 500 mM NaCl and 0.02% Tween 20 for 10 min three times and rinse with water.
13. Dry the slide by purging with Ar gas and incubate the antibody-treated slide with Cy5-labeled goat anti-antibody (10 µg/ml) in 10 mM sodium phosphate (pH 7.2) containing 500 mM NaCl and 0.02% Tween 20 for 1.5 h at room temperature.
14. Wash the slide with 10 mM sodium phosphate (pH 7.2) containing 500 mM NaCl and 0.02% Tween 20 for 10 min three times and rinse with water.
15. Dry the slide by purging with Ar gas and detect the bound antibody using a fluorescence scanner.

## 4. Notes

1. High concentrations of proteins sometimes result in the precipitation of proteins. In this case, decrease the concentration of proteins in your sample.
2. Addition of glycerol suppresses the undesired evaporation of nanodroplets during spotting and immobilization.
3. The addition of organic solvent (in this case, glycerol) to the buffer causes a change of pH. Therefore, the pH value of the buffer after the addition of an organic solvent should be adjusted to the desired value.

4. Quality of the modified glass slides is very important for reproducible results. It is recommended that those with little experience purchase the modified glass slides from commercial suppliers.

5. Pro-Q Emerald 300 stain has an excitation maximum at approximately 280 nm and an emission maximum at 530 nm.

6. Gel- and blot-staining images can be documented using a photographic camera or CCD camera. Appropriate filters are essential to obtain the maximum sensitivity.

7. With a UV light box, the protein spots and bands of interest can be excised for mass spectrometric analysis.

8. To prevent the clogging of the CE, LC columns, or nanospray needles, all solutions should be filtered before injecting into CE or LC/MS. For filtering, the 0.2 μm membranes are recommended.

9. An important issue for CE/MS is the formation of a closed electric circuit with appropriate ground connection. The electrical scheme of the circuit in **Fig. 1** should be followed closely.

10. A violet spot at the needle tip during an online sheathless CE/MS experiment indicates electrical (Corona) discharge, which is prone to occur in particular under sheathless unpressurized reverse polarity CE/MS. If such an event occurs, the ESI and the counterelectrode voltages are to be stepwise decreased.

11. The metal deposition of the nanospray needles is degradable and cannot be refurbished. Degradation in time of the metal coat results in the interruption of the electrical contact and deterioration of the circuit performance. Therefore, changing the needle after 5–6 experiments is recommended.

12. For the non-pressurized reverse-polarity CE/(–) nanoESI MS mode, the use of the formic acid/ammonia buffer at pH value 2.8 in combination with uncoated internal capillary walls, results in almost total suppression of the electroosmotic flow. Under these conditions, the separation and migration of the negatively charged analytes toward the anode will be driven only by the electrical force; therefore the migration time might be longer.

13. Blocking of unreacted thiol groups with *N*-ethylmaleimide prevents oxidative disulfide-bond formation between surface thiols and cysteine residues of proteins during incubation with labeled proteins.

14. Quenching of the unreacted epoxide groups on the surface by treatment with glycine or 2-aminoethanol is not necessary. Incubation of the carbohydrate microarrays with BSA before probing with proteins or cells quenches the unreacted epoxide groups.

15. Quenching of the reaction by addition of 2-aminoethanol to remove the unreacted FITC is not necessary. Free dye is easily removed by gel filtration.

16. Unreacted dye can be also removed from conjugated proteins by dialysis, although this is not as efficient or rapid as gel filtration. Therefore, it is recommended that isolation of conjugated proteins be achieved by gel filtration.

17. SYTO 83 is more efficient than propidium iodide for staining live cells. However, propidium iodide is suitable for staining dead cells.

18. Unseparated stained cells give a better binding to carbohydrates immobilized on the surface than the stained cells collected by centrifugation followed by suspension of cells with buffer.

19. One of the biggest problems for microarray experiments is the non-specific interaction of the probing proteins with modified surfaces. Treatment of the microarrays with BSA considerably decreases these non-specific interactions. Hydrophilically modified surfaces such as poly(ethylene glycol) (PEG) considerably suppress these undesired interactions even without treatment with BSA.

20. Tween 20 should be added to a solution of dye-labeled proteins before use. Long-term storage of a protein solution containing Tween 20 decreases protein activity.

## Acknowledgments

Financial support from NRL program (KOSEF/MOST, Dr. Injae Shin), USA National Cancer Institute R21 CA111949-01 (Dr. Bin Ye) as well as German DFG/SFB 492 and Romanian CEx. 14 (Dr. Alina D. Zamfir) are acknowledged and greatly appreciated.

## References

1. Morelle, W. and Michalski, J. C. (2005) Glycomics and mass spectrometry *Curr Pharm Des* **11,** 2615–45.
2. Peter-Katalinic, J. (2005) Methods in enzymology: O-glycosylation of proteins *Methods Enzymol* **405,** 139–71.
3. Blixt, O., Head, S., Mondala, T., Scanlan, C., Huflejt, M. E., Alvarez, R., Bryan, M. C., Fazio, F., Calarese, D., Stevens, J., Razi, N., Stevens, D. J., Skehel, J. J., van Die, I., Burton, D. R., Wilson, I. A., Cummings, R., Bovin, N., Wong, C. H., and Paulson, J. C. (2004) Printed covalent glycan array for ligand profiling of diverse glycan binding proteins *Proc Natl Acad Sci USA* **101,** 17033–8.
4. Disney, M. D. and Seeberger, P. H. (2004) The use of carbohydrate microarrays to study carbohydrate-cell interactions and to detect pathogens *Chem Biol* **11,** 1701–7.
5. Fukui, S., Feizi, T., Galustian, C., Lawson, A. M., and Chai, W. (2002) Oligosaccharide microarrays for high-throughput detection and specificity assignments of carbohydrate-protein interactions *Nat Biotechnol* **20,** 1011–7.
6. Houseman, B. T. and Mrksich, M. (2002) Carbohydrate arrays for the evaluation of protein binding and enzymatic modification *Chem Biol* **9,** 443–54.
7. Kawahashi, Y., Doi, N., Takashima, H., Tsuda, C., Oishi, Y., Oyama, R., Yonezawa, M., Miyamoto-Sato, E., and Yanagawa, H. (2003) In vitro protein microarrays for detecting protein-protein interactions: application of a new method for fluorescence labeling of proteins *Proteomics* **3,** 1236–43.
8. Kohn, M., Wacker, R., Peters, C., Schroder, H., Soulere, L., Breinbauer, R., Niemeyer, C. M., and Waldmann, H. (2003) Staudinger ligation: a new

immobilization strategy for the preparation of small-molecule arrays. *Angew Chem Int Ed Engl* **42,** 5830–4.

9. Lee, M. R. and Shin, I. (2005) Fabrication of chemical microarrays by efficient immobilization of hydrazide-linked substances on epoxide-coated glass surfaces. *Angew Chem Int Ed Engl* **44,** 2881–4.

10. Lee, M. R. and Shin, I. (2005) Facile preparation of carbohydrate microarrays by site-specific, covalent immobilization of unmodified carbohydrates on hydrazide-coated glass slides *Org Lett* **7,** 4269–72.

11. Nimrichter, L., Gargir, A., Gortler, M., Altstock, R. T., Shtevi, A., Weisshaus, O., Fire, E., Dotan, N., and Schnaar, R. L. (2004) Intact cell adhesion to glycan microarrays *Glycobiology* **14,** 197–203.

12. Park, S., Lee, M. R., Pyo, S. J., and Shin, I. (2004) Carbohydrate chips for studying high-throughput carbohydrate-protein interactions *J Am Chem Soc* **126,** 4812–9.

13. Park, S. and Shin, I. (2002) Fabrication of carbohydrate chips for studying protein-carbohydrate interactions *Angew Chem Int Ed Engl* **41,** 3180–2.

14. Ratner, D. M., Adams, E. W., Su, J., O'Keefe, B. R., Mrksich, M., and Seeberger, P. H. (2004) Probing protein-carbohydrate interactions with microarrays of synthetic oligosaccharides *ChemBioChem* **5,** 379–82.

15. Schwarz, M., Spector, L., Gargir, A., Shtevi, A., Gortler, M., Altstock, R. T., Dukler, A. A., and Dotan, N. (2003) A new kind of carbohydrate array, its use for profiling antiglycan antibodies, and the discovery of a novel human cellulose-binding antibody *Glycobiology* **13,** 749–54.

16. Shin, I., Cho, J. W., and Boo, D. W. (2004) Carbohydrate arrays for functional studies of carbohydrates *Comb Chem High Throughput Screening* **7,** 565–74.

17. Shin, I., Park, S., and Lee, M. R. (2005) Carbohydrate microarrays: an advanced technology for functional studies of glycans *Chem Eur J* **11,** 2894–901.

18. Wang, D., Liu, S., Trummer, B. J., Deng, C., and Wang, A. (2002) Carbohydrate microarrays for the recognition of cross-reactive molecular markers of microbes and host cells *Nat Biotechnol* **20,** 275–81.

19. Kitamura, N., Guo, S., Sato, T., Hiraizumi, S., Taka, J., Ikekita, M., Sawada, S., Fujisawa, H., and Furukawa, K. (2003) Prognostic significance of reduced expression of beta-N-acetylgalactosaminylated N-linked oligosaccharides in human breast cancer *Int J Cancer* **105,** 533–41.

20. Ye, B., Skates, S., Mok, S. C., Horick, N. K., Rosenberg, H. F., Vitonis, A., Edwards, D., Sluss, P., Han, W. K., Berkowitz, R. S., and Cramer, D. W. (2006) Proteomic-based discovery and characterization of glycosylated eosinophil-derived neurotoxin and COOH-terminal osteopontin fragments for ovarian cancer in urine *Clin Cancer Res* **12,** 432–41.

21. Mechref, Y., Muzikar, J., and Novotny, M. V. (2005) Comprehensive assessment of N-glycans derived from a murine monoclonal antibody: a case for multimethod-ological approach *Electrophoresis* **26,** 2034–46.

22. Bindila, L., Peter-Katalinic, J., and Zamfir, A. (2005) Sheathless reverse-polarity capillary electrophoresis-electrospray-mass spectrometry for analysis of underivatized glycoconjugates *Electrophoresis* **26,** 1488–99.

23. Zamfir, A. and Peter-Katalinic, J. (2001) Glycoscreening by on-line sheathless capillary electrophoresis/electrospray ionization-quadrupole time of flight-tandem mass spectrometry *Electrophoresis* **22,** 2448–57.
24. Zamfir, A., and Peter-Katalinic, J. (2004) Capillary electrophoresis-mass spectrometry for glycoscreening in biomedical research *Electrophoresis* **25,** 1949–63.

# 3

# Surface-Enhanced Laser Desorption/Ionization Mass Spectrometry for Protein and Peptide Profiling of Body Fluids

Audrey Gagnon, Qian Shi, and Bin Ye

## Summary

Recently, Ciphergen Biosystems (Fremont, CA, USA) has developed a technique called surface-enhanced laser desorption/ionization (SELDI). This technology is based on ProteinChips® with chemically or biochemically modified surfaces for the selective retention and enrichment of protein subsets from a complex protein mixture. The proteins or peptides of interest can then be identified by using mass spectrometry (SELDI-MS). This highly sensitive and high-throughput technique can detect minute differences in proteins and peptide profiles between biological samples. The versatility of this technology enables a wide variety of applications in basic research, clinical, proteomic, and drug discovery such as identification of novel biomarkers associated with certain diseases or treatments. Many disease biomarkers or biomarker patterns have been identified using SELDI-MS in various laboratories. In this chapter, we provide a general guide to the profiling of proteins or peptides as well as biomarker discovery using the most common body fluids such as serum/plasma, nipple aspiration fluid, and urine.

**Key Words:** Protein mixture; body fluids; SELDI-TOF-MS; protein fractionation; profiling; disease biomarker; serum; plasma, nipple aspiration fluid; urine.

From: *Methods in Molecular Biology, vol. 441: Tissue Proteomics: Pathways, Biomarkers, and Drug Discovery*
Edited by: B.C.-S. Liu © Humana Press, Totowa, NJ

## 1. Introduction

Compared to the genome, the proteome provides a more realistic view of a cell's biological status. Indeed, the DNA content of a cell is static and does not change over time or in response to treatment. Therefore, analysis of the human proteome is expected to be more useful than gene analysis for evaluating the presence or progression of a disease or to monitor a treatment response. Thus, proteomics technologies can bridge the gap between the genome sequence and the cellular behaviors as they assess components that reflect more a cell's biological status *(1)*.

Studies have shown that the serum proteome contains many unknown proteins with low molecular weight (below 15 kDa), below the detection sensitivity of conventional 2D gels *(2)*. However, SELDI-MS can be used to detect a few femtomoles of a specific protein from a complex mixture and determine its molecular weight with an error rate of less than 0.2% *(3)*. Thus, very complex body fluids such as serum or urine can be applied to the SELDI-MS ProteinChip® arrays for high-sensitivity fractionation and profiling. Proteins with a specific chemical charge, hydrophobic proteins, metal-binding proteins, or proteins with a certain affinity for a previous coated target (such as an antibody or a receptor), bind selectively to the ProteinChip® surface, whereas the contaminating background is washed away in a process called retention mapping. Then, an energy-absorbing molecule such as trifluoroacetic acid (TFA) or α-cyano-4-hydroxycinnamic acid (CHCA) is applied to the ProteinChip® arrays, and the sample is analyzed directly by using laser desorption/ionization-time of flight mass spectrometry (SELDI-TOF-MS). Measurements of proteins and peptides are transformed into the mass-to-charge ratio (m/z). SELDI-TOF-MS analysis allows the rapid, sensitive, and high-resolution screening of 8–24 samples on a single array.

The usefulness of this technology has been proven by its wide application and frequent publications in recent years. The main advantages of the SELDI-TOF-MS ProteinChip® technology include:

1. Ease of use. Indeed, in most cases, users do not have to be specifically trained in a mass spectrometry laboratory.
2. High sensitivity of detection.
3. Easily adaptable for complex biological specimens such as blood, serum, plasma, and urine.
4. An array surface that provides selective chemical and biochemical affinity for retention of proteins, peptides, and various bio-molecules without traditional chromatographic purification procedures.

In this chapter, we summarize the most recent methods and protocols for the SELDI-TOF-MS ProteinChip® arrays platform in protein profiling and biomarker discovery using serum, plasma, nipple aspiration fluid (NAF), and urine.

## 2. Materials

### 2.1. Serum Profiling *(see Note 1)*

Serum is one of the most common clinical samples in the world. Indeed, a large number of clinical tests are conducted using plasma/serum samples, as they contain many of molecules indicative of diseases. Serum is also the most common sample used for biomarker discovery for various reasons: it is easily obtained, almost all clotting factors are removed, and it remains a rich source of molecules indicative of systemic functions *(4)*. Marker screening technology, such as SELDI-TOF-MS, is widely used on plasma/serum samples to identify sets of biomarkers for clinical use, such as disease detection, therapy monitoring, and prognosis.

The reference value of total protein in serum is 6–8 g/dL, in which albumin constitutes about 70% and immunoglobulins (Igs) 20%. The total protein content is relatively stable within a person and within a population, with the exception of diseases or conditions that diminish the total protein content, such as malnutrition, liver disease, certain types of renal disease, and rare conditions such as hyperlipidemia *(5)*. In most research fields, these conditions are not a significant factor, and the total proteins in the serum can be assumed to be steady/unchanging, thus eliminating the total protein normalization step. In SELDI-TOF-MS, the normalization of total proteins can be partially achieved after data acquisition using the total ion current normalization tool.

1. Bio-processor (Ciphergen Biosystems)
2. One of the following chip types: Metal-binding IMAC3 chips (Ciphergen Biosystems); Weak cation exchange CM10 chips (Ciphergen Biosystems); Hydrophobic H4 chips (Ciphergen Biosystems); Weak cation exchange WCX2 chips (Ciphergen Biosystems); or, Strong anion exchange chips SAX2 (Ciphergen Biosystems), and their relevant updated chips
3. 96-well plates (Nalge Nunc, Naperville, IL, USA)
4. MS1 Mini shaker (IKA, Wilmington, NC, USA)
5. Denaturation buffer: 9 M urea, 2% 3-[(3-cholamidopropyl) dimethylammonio]-1 propanesulfonate (CHAPS), 50 mM Tris–HCl (pH 9.0)
6. Binding buffers: 0.1 M sodium phosphate buffer (pH 7.0), 0.5 M NaCl, 10 mM imidazol (*see* **Note 2**) (see **Table 1** for a list of the different binding/washing buffers that can be used)
7. 100 mM $CuSO_4$
8. Saturated sinapinic acid (SPA or 3,5-dimethoxy-4-hydroxycinnamic acid) solution: 12.5 mg/ml SPA in 50% acetonitrile and 0.1% TFA (*see* **Note 4**).

**Table 1**
**Summary of Protocols for Serum**

| Chip type | Chip preparation | Binding buffer | Washes | Comments | Reference |
|---|---|---|---|---|---|
| IMAC3, Metal binding (most widely used for serum profiling) | 100 mM $CuSO_4$ humid chamber 15 min, neutralize 5 min with 0.1 M NaOAc pH 4.0 | 0.1 M sodium phosphate buffer (pH 7.0), 0.5 M NaCl, 10 mM imidazol, pH 7.0 (*see* **Note 2**) | Three times, 5 min with binding buffer, once with water | Can also be loaded with nickel or gallium (no equilibration needed) | *(23,24)* |
| CM10, cation exchange | 10 mM HCl, 5 min | 20 mM $Na_2HPO_4$, pH 5.0 or 100 mM NaOAc + 30 mM NaCl pH 4.0 | Twice, 5 min with binding buffer, twice with water | 0.1 M NaOAc ph 4 can also be used as a binding/washing buffer (*see* **Note 3**) | *(1,25,26)* |
| H4, hydrophobic | 10% acetonitrile, 5 min | 50 mM HEPES pH 7.0 + 10% acetonitrile + 0.1% TFA | Three times, 5 min with binding buffer, once with water | Can also use 50% acetonitrile + 0.1% trifluoroacetic acid + 0.5 M NaCl as a binding/washing buffer | *(26)* |
| WCX2, Weak cation exchange | 10 mM HCl, 5 min | 0.1 M NaOAc pH4 (low stringency) | Twice 5 min with binding buffer, Once with water | A high-stringency buffer can also be used, 50 mM HEPES, pH 7, 100 mM ammonium acetate can also be used | *(27,28)* |
| SAX2, Strong anion exchange | Neutralization with 100 µl of 100 mM ammonium acetate pH 6.0, 5 min | 0.05M Tris + 0.1% triton X-100, pH 9 | Three times 5 min each with binding buffer, Twice with water | 10–100 mM Tris–HCl pH 7.5–9.0 or 10–100 mM ammonium acetate pH 4.0–6.0 can also be used as binding/washing buffer | *(29)* |

## 2.2. NAF Protein Profiling

The human mammary gland is composed of ductal–alveolar systems that originate at the nipple and branch toward the chest wall. It is believed that most breast carcinomas (70–80%) arise from the epithelial cells lining these ducts. As the breast epithelium secretes fluids in the luminal compartment of the gland, it is thought that NAF would be useful in the study of breast diseases, such as breast cancer *(6)*. Obtaining NAF is non-invasive and cheap, and it yields a small set of breast-specific proteins. Because these proteins are secreted, they are the final processed form of the protein, which makes proteomics analyses less ambiguous. Moreover, they can provide us with information about changes in protein translation rates, post-translational modifications, sequestration, or degradation that might lead to disease. However, the procedure is less standardized than blood collection, and its validation requires the demonstration that biological markers in the fluid correlate with breast tissue pathology *(7)*.

1. Denaturation buffer: 9 M urea, 1.2% CHAPS, 50 mM Tris–HCI (pH 9.0)
3. Bio-processor
4. One of the following chip types: Metal-binding IMAC3 chips; Normal phase NP10 chips; Hydrophobic H4 chips; or, Weak cation exchange WCX2 chip
5. 96-well plates
6. MS1 mini shaker

## 2.3. Urinary Protein/Peptide Profiling

In contrast to serum, urine can fluctuate dramatically in protein and salt concentrations, and both the inter- and intra-patient variability can be quite large. Typically, the urine samples containing the highest protein concentrations are first and second morning void samples. To minimize the impact of this variability in the clinical setting, samples are often collected over the course of 24 h *(8)*. Urine from normal individuals has low protein concentrations (150 mg/day). Approximately 10% of this protein is albumin; the remaining proteins are various serum proteins as well some that arise from the renal cells themselves *(9)*. Fifty percent of the protein is the Tamm–Horsfall mucoprotein (96 kDa), which is secreted by cells of the ascending limb of the loop of Henle *(10)*.

High amounts of protein are themselves generally indicative of a pathologic condition. The nephrotic syndrome (proteinuria, edema, and hyperlipidemia) is one of many well-characterized clinical conditions giving rise to proteinuria. In some situations, the specific types of proteins found in the urine can give hints as to pathology. Damage to the renal tubules generally results in an increase in proteins of low molecular weight (such as beta2-microglobulin, 11.6 kDa;

lysozyme, 14 kDa; and immunoglobulin light chain, 22 kDa) but not in proteins of higher molecular weight (such as albumin). This situation generally does not lead to the clinical picture of the nephrotic syndrome, which is instead usually caused by glomerular disease. Finally, extra-renal processes such as urinary tract infection can also lead to proteinuria *(11)*.

As with all clinical samples, urine samples for profiling should be obtained in a uniform manner. When looking at the sample, be aware of variability in cellularity and color (such as what might result from hemolysis) *(11)*. Moreover, aliquoting the samples is critical, as repeated freeze-thaw cycles will generally lead to degradation of proteins, particularly when the overall concentration of protein in the urine is low. However, it has been shown that serum or urine samples could sustain up to four freeze-thaw cycles before the protein profile is significantly changed *(12)*. Generally, denaturing of the samples will lead to more reproducible results. Our recommendation is to take an initial volume of sample, centrifuge, and then add denaturing buffer in the indicated ratio. The denatured sample, which is more stable than the non-denatured sample, can then be aliquoted and frozen at −80 °C. However, because urine is less complex than serum or plasma, fractionation is not usually required. However, size fractionation is one option that may be useful to isolate smaller proteins.

1. Denaturation buffer: 9 M urea, 1.2% CHAPS, 50 mM Tris–HCI (pH 9.0)
2. Bio-processor
3. One of the following chip types: Metal-binding IMAC3 chips; Normal phase NP10 chips; Hydrophobic H4 chips; Weak cation exchange WCX2 chip 9; or, Normal phase NP20 (Ciphergen Biosystems)
4. 96-well plates
5. MS1 mini shaker

## 3. Methods

### *3.1. Serum Protein Profiling (see Note 5)*

The following protocol is an adaptation of the Ciphergen Biosystems, Inc. user manual and of different published research papers. The Ciphergen Biosystems protocol for serum proteins profiling begins with an anion exchange fractionation step that uses step-wise pH elutions. A total of seven fractions (pH 3.0–9.0) are collected for each sample. Each fraction is bound to a ProteinChip®. However, because very few published articles use this fractionation step without any problems or variability (*see* **Note 6**), we will not present the fractionation protocol in this chapter. For a summary of available protocols for serum protein profiling, see **Table 1**.

### 3.1.1. Sample Denaturation

1. Aliquot 5 µL of serum to each well of a 96-well plate
2. Add 5 µL of denaturation buffer to each sample
3. Vortex 30 min (*see* **Note 7**).
4. Add 40 µL of denaturation buffer to 10 µL of the denatured serum samples (*see* **Note 8**).

### 3.1.2. Chip Loading, Neutralization, Equilibration, and Binding

1. Place chips in a bio-processor
2. Load IMAC3 chips with copper (refer to **Table 1** for other metals that can be used). Load 50 µL of 100 mM $CuSO_4$ onto each spot, then incubate 15 min in a humid chamber. Remove the $CuSO_4$ and rinse with water once.
3. Neutralize all chips. Load 50 µL of the appropriate binding buffer onto each spot. Vortex for 5 min, then remove the binding buffer. Rinse with water once.
4. To equilibrate chips, add 100 µL of PBS into each well. Vortex for 5 min, then remove buffer. Repeat this once.
5. Bind the proteins to chips by adding 50 µl of denatured/diluted serum sample into each well. Vortex for 30 min, then remove sample and buffer.
6. Wash chips by adding 200 µl of the appropriate binding buffer. Vortex for 5 min and repeat wash once. Rinse with water once.

### 3.1.3. Matrix Addition

1. Remove chips from bio-processor
2. Air-dry the chips
3. Add 0.5 µl of saturated SPA solution in 50% acetonitrile and 0.5% TFA
4. Air-dry

### 3.1.4. Scanning, mass Spectrum Data Collection, and Calibration

The chips were scanned using a PBSII-C instrument (Ciphergen Biosystems) equipped with an autoloader. SELDI software version 3.1 was used to control the parameters of data collection. The parameter settings were: a 100,000 kDa high mass collection limit, and optimization ranges from 3000 to 50,000 kDa with focus at 5000 kDa. The laser intensity used was optimized for each fraction. An automatic protocol was implemented to collect the data from each set of chips. The laser intensity was set to 200, 230, or 250 depending on the experiment; the detection sensitivity was set at 10; and the resulting data collected were averaged and stored as the spectrum for each spot. The molecular weight accuracy of the system was calibrated with a human recombinant insulin control chip (5807.7 kDa) (Ciphergen Biosystems) immediately before scanning.

### *3.2. NAF Protein*

For a summary of available protocols for NAF profiling, *see* **Table 2**.

#### *3.2.1. Sample Collection and Treatment*

NAF can be collected in a capillary tube with the help of a breast pump. The volume of the NAF can be measured directly from the capillary tube. All the NAF samples should be coded to allow blind analysis. The capillary tubes are snap-frozen in liquid nitrogen, placed in 1.5-ml tubes, labeled, and stored immediately in liquid nitrogen *(7)*.

Because NAF normally has a less complex proteome than serum or plasma, there is no need to fractionate it before proceeding with SELDI-TOF-MS. NAFs are denatured using buffer at a ratio of 1 µl NAF to 9 µl of buffer. The procedure

**Table 2**
**Summary of Protocols for Nipple Aspiration Fluid**

| Chip type | Chip preparation | Binding buffer | Washing buffer | Comments | Reference |
| --- | --- | --- | --- | --- | --- |
| IMAC3, Metal binding | 100 mM $CuSO_4$ | 0.1 M sodium phosphate buffer (pH 7.0), 0.5 M NaCl, 10 mM imidazol | Twice with 1× PBS, Twice with water | 1 µg of breast fluid proteins were used on the chip | *(6)* |
| H4, Hydrophobic | 10% acetonitrile | 50 mM HEPES, pH 7 | Three times with 10% acetonitrile, Three times with water | | *(13)* |
| SAX2, Strong anion exchange | None | 100 mM Tris–HCl, pH 8 + 50 mM NaCl + 0.1% OG | Twice with binding buffer, Three times with water | NAF was diluted 100-fold before analysis | *(13)* |
| NP10, Normal phase | none | 100 mM Tris–HCl, pH 8 + 50mM NaCl + 0.1% OG | Twice with binding buffer, Three times with water | NAF was diluted 100-fold before analysis | *(13)* |

NAF, nipple aspiration fluid.

outlined below utilizes a metal rod to crush the capillary tubes. This avoids possible cross-contamination and can be easily adapted for high-throughput studies *(13)*.

1. Thaw the NAF samples on ice
2. Put capillary with NAF in a 1.5-ml tube containing 100 µl of 100 mM Tris–HCl, pH 8.0
3. Crush the capillary with a metal or glass rod
4. Vortex tubes to disperse the sample
5. Spin the sample 5 min at 14,000 × *g*
6. Transfer supernatant to a clean tube
7. Determine protein concentration with Pierce BCA assay (Pierce Biotechnology, Rockford, IL, USA)
8. Dilute NAF proteins to 3.6 µg/µl

### 3.2.2. Chip-Binding Protocol

The chip loading, neutralization, equilibration, binding, and washing as well as the matrix addition and scanning procedures are the same as **Subheadings 3.1.2**, **3.1.3**, and **3.1.4**. Refer to Table 2 for reagents.

### 3.3. Urine Protein and Peptide Profiling

For a summary of available protocols for urine protein profiling, *see* **Table 3**.

### 3.3.1. Sample Treatment

As mentioned previously, we recommend denaturing the samples before using them for SELDI. Also, depending on the expected protein concentration, the samples can be put directly on the chip or diluted in denaturation buffer before use.

### 3.3.2. H4 Arrays

1. Place chips into bio-processor
2. Add 100 µl of 50% acetonitrile, incubate 5 min
3. Remove acetonitrile solution.

### 3.3.3. All Chips

#### 3.3.3.1. Equilibrate Chips

1. Add 150 µl of the appropriate binding buffer to each well
2. Vortex 5 min
3. Remove buffer
4. Repeat equilibration once

**Table 3**
**Summary of Protocols for Urine**

| Chip type | Chip preparation | Binding buffer | Washing buffer | Comments | References |
|---|---|---|---|---|---|
| IMAC3, Metal binding | 100 mM CuSO$_4$, 5 min in a humid chamber, neutralization with 100 mM NaOAc pH 4.0 for 5 min | None | Four times, 5 min with 1× phosphate-buffered saline (PBS) | | *(30,31)* |
| CM10, Cation exchange | 100 mM HCl, 5 min | 0.1 M NaOAc, pH 4.0 | Twice, 5 min with binding buffer, Once with water | Urine protein was adjusted to 50 µg/ml | *(25)* |
| NP20, Normal phase | None | None | Three times, 5 min with water | CHCA 35% in 50% acetonitrile and 0.5% trifluoroacetic acid works better than sinapinic acid | *(32,33)* |
| H4, Hydrophobic | 50% acetonitrile, 5 min | Samples are diluted in acetonitrile | 20% acetonitrile, 5 min | Washes were done in 50% acetronitrile (increased stringency) | *(31,34)* |
| SAX2, Anion exchange | 10 mM borate buffer pH 9.0, 5 min | 10 mM borate buffer pH 9.0 | 10 mM borate buffer pH 9.0, 5 min, once with water | | *(34)* |
| WCX2, Weak cation exchange | 10 mM HCl, 5 min | 100 mM ammonium acetate, pH 6.5 | Twice with binding buffer | 5 µl of non-diluted urine were applied to the chips | *(35)* |

### 3.3.3.2. 2-Step Binding of Urine Sample to Arrays

1. Add 175 µl of the appropriate binding buffer to each well
2. Add 25 µl of denatured urine—return sample plate to ice
3. Vortex 30 min
4. Remove sample and buffer
5. Repeat binding once

### 3.3.3.3. Chips Washing

1. Add 100 µl of the appropriate washing buffer to each well
2. Vortex 5 min
3. Remove buffer
4. Repeat the wash twice
5. Rinse with 100 µL of water (no water wash to H50 arrays)

### 3.3.3.4. Matrix Addition

1. Add 2× 0.5µl of saturated SPA solution made in 50% acetonitrile and 0.5% TFA
2. Air-dry between each application

### 3.3.3.5. Scanning, Mass Spectrum Data Collection, and Calibration

The procedure for calibration, scanning, and data acquisition is the same as that outlined in **Subheading 3.1.4**.

## 3.4. Other Body Fluids and Relevant Methods

### 3.4.1. Cervical and Amniotic Fluids

To monitor certain gynecologic conditions, amniotic and cervical fluids can be useful, especially the latter, which are collected using a non-invasive procedure. To assess whether cervical and amniotic fluids can be studied with SELDI-TOF-MS, Ruetschi et al., applied these samples to the strong anion exchange Q10 Proteinchip®. Briefly, the fluids were not denatured but diluted 1:4 and then applied to the Q10 chip (binding buffer, 100 mM Tris–HCl, pH 9.0). Their study indicated a great deal of individual variation in the amniotic fluid composition. Additionally, when cervical fluid was analyzed using the same technique, no difference was noted between cases and controls. Thus, the use of cervical and amniotic fluids to monitor gynecologic diseases could depend on the disease itself. The procedure would then have to be adapted and optimized for each experimental design *(14)*.

### 3.4.2. Saliva

There is an increasing interest in using saliva to diagnose systemic diseases because its collection is very simple *(15)*, safe, inexpensive, and causes minimal discomfort to the patient. More importantly, saliva contains proteins and peptides that are frequently altered in the presence of a systemic disease *(15–17)*; however, the levels of certain markers in saliva are not always a reliable reflection of levels in the serum. For example, lipophilic molecules diffuse more easily into saliva than lipophobic molecules. Also, salivary composition can be affected by the method of collection and the degree of stimulation of salivary flow. Finally, saliva also contains proteolytic enzymes derived from the host's oral microorganisms *(17)*. These enzymes can affect the stability of certain diagnostic markers. Nevertheless, saliva could be very useful for qualitative assessment (presence or absence) of a marker. Recently, Streckfus et al. performed a feasibility study using saliva on three different chips (hydrophobic H4, strong anion exchange SAX2, and weak cation exchange WCX2) and different binding buffer pHs. They found that it is possible to assay salivary specimens by using SELDI-TOF-MS. Moreover, they showed that saliva does not require special preparation and that the WCX2 (weak cation exchange, binding buffer pH 3.5 instead of pH 4) Proteinchip® gives much lower noise signals *(18)*.

### 3.4.3. Cerebrospinal Fluid

Cerebrospinal fluid (CSF) constantly circulates through cavities in the brain and spinal cord. Thus, it is thought to contain peptides and proteins that play critical roles in many physical processes and to reflect the pathological state of the brain tissues and other parts of the central nervous system *(19,20)*. Recently, Guerreiro et al. used denatured and native CSF samples on four different chip surfaces: weak cation exchange CM10, strong anion exchange Q10, hydrophobic H50, and metal-binding IMAC3. They observed many peaks with all chips used. Their study also showed that a greater number of resolved peaks are obtained using denatured samples as well as non-reducing conditions. They also found that more peaks were obtained using a binding buffer without triton X-100 (*see* **Table 1**) *(21)*.

### 3.5. Conclusion

In conclusion of this protocol, we have found that the SELDI-TOF-MS technique is a very powerful tool to decipher a variety of complex body fluids. Furthermore, many chips' surfaces (strong anion exchange, weak cation exchange, hydrophobic, metal binding, normal phase, etc.) can be used for profiling commonly studied body fluids such as serum/plasma, NAF, and urine. Moreover, each chip provides different but complementary information about the composition and the profile of the samples studied. Thus, SELDI-TOF-MS has many uses and is very helpful in the study of complex mixtures of proteins such as body fluids.

## 4. Notes

1. All reagents, powders, and water used in the protocols should be high-performance liquid chromatography (HPLC) grade.
2. Increasing the concentration of imidazol will increase the selectivity of the surface.
3. Choose a binding buffer that is at least 1 pH unit below the PI of the target protein.
4. SPA and TFA are toxic if ingested or inhaled and require proper protection (safety glasses as well as an adequately ventilated room). For the saturated SPA solution: make a solution of 12.5 mg/ml of SPA in 0.5% TFA and 50% acetonitrile. Vortex 2 min, let stand for 5 min at room temperature, spin 10 min at maximum speed to remove undissolved SPA. Aliquot the supernatant in small volumes and store them at $-80\,°C$. The solution is then stable for 1 month.
5. All steps are carried out at room temperature unless stated otherwise.
6. We recommend that the serum samples be albumin- and IgG-depleted before they are used for SELDI-TOF-MS. This removes the two most abundant proteins in the serum and greatly reduces the noise background. However, albumin and IgGs can also act as carriers, and peptides might be lost during the depletion process *(22)*. That makes it the users' decision whether to deplete the serum samples.
7. A good vortex (plate shaker) is crucial for all protocols (rpm 250) for optimal binding and washing of the chips.
8. A starting serum dilution of 1:10 is suggested. However, the dilution will have to be adjusted according to each type of experiment.

## Acknowledgments

Financial support from the NIH Cancer Institute (R21 CA11949-01) is acknowledged. We also acknowledge Drs. Daniel Cramer, Samuel Mok and Brian Liu from Brigham and Women's Hospital and Dr. Miron Alexander from Dana-Farber cancer Institute for their encouraging discussion during the manuscript preparation.

## References

1. Engwegen, J. Y., Gast, M. C., Schellens, J. H., and Beijnen, J. H. Clinical proteomics: searching for better tumour markers with SELDI-TOF mass spectrometry. *Trends Pharmacol Sci, 27:* 251–259, 2006.
2. Simpkins, F., Czechowicz, J. A., Liotta, L., and Kohn, E. C. SELDI-TOF mass spectrometry for cancer biomarker discovery and serum proteomic diagnostics. *Pharmacogenomics, 6:* 647–653, 2005.

3. Austen, B. M., Frears, E. R., and Davies, H. The use of seldi proteinchip arrays to monitor production of Alzheimer's betaamyloid in transfected cells. *J Pept Sci, 6:* 459–469, 2000.

4. Banks, R. E., Stanley, A. J., Cairns, D. A., Barrett, J. H., Clarke, P., Thompson, D., and Selby, P. J. Influences of blood sample processing on low-molecular-weight proteome identified by surface-enhanced laser desorption/ionization mass spectrometry. *Clin Chem, 51:* 1637–1649, 2005.

5. Hortin, G. L. The maldi-tof mass spectrometric view of the plasma proteome and peptidome. *Clin Chem, 52:* 1223–1237, 2006.

6. Li, J., Zhao, J., Yu, X., Lange, J., Kuerer, H., Krishnamurthy, S., Schilling, E., Khan, S. A., Sukumar, S., and Chan, D. W. Identification of biomarkers for breast cancer in nipple aspiration and ductal lavage fluid. *Clin Cancer Res, 11:* 8312–8320, 2005.

7. Alexander, H., Stegner, A. L., Wagner-Mann, C., Du Bois, G. C., Alexander, S., and Sauter, E. R. Proteomic analysis to identify breast cancer biomarkers in nipple aspirate fluid. *Clin Cancer Res, 10:* 7500–7510, 2004.

8. de Jong, P. E. and Gansevoort, R. T. Screening techniques for detecting chronic kidney disease. *Curr Opin Nephrol Hypertens, 14:* 567–572, 2005.

9. Mathieson, P. W. The cellular basis of albuminuria. *Clin Sci (Lond), 107:* 533–538, 2004.

10. Ruzhanskaya, A. V., Milenina, O. E., Kravtsov, E. G., Dalin, M. V., and Gabrielyan, N. I. Methodological approaches to detection of Tamm-Horsfall protein. *Bull Exp Biol Med, 140:* 330–333, 2005.

11. Simerville, J. A., Maxted, W. C., and Pahira, J. J. Urinalysis: A comprehensive review. *Am Fam Physician, 71:* 1153–1162, 2005.

12. Traum, A. Z., Wells, M. P., Aivado, M., Libermann, T. A., Ramoni, M. F., and Schachter, A. D. SELDI-TOF MS of quadruplicate urine and serum samples to evaluate changes related to storage conditions. *Proteomics, 6:* 1676–1680, 2006.

13. Sauter, E. R., Shan, S., Hewett, J. E., Speckman, P., and Du Bois, G. C. Proteomic analysis of nipple aspirate fluid using SELDI-TOF-MS. *Int J Cancer, 114:* 791–796, 2005.

14. Ruetschi, U., Rosen, A., Karlsson, G., Zetterberg, H., Rymo, L., Hagberg, H., and Jacobsson, B. Proteomic analysis using protein chips to detect biomarkers in cervical and amniotic fluid in women with intra-amniotic inflammation. *J Proteome Res, 4:* 2236–2242, 2005.

15. Streckfus, C. F. and Bigler, L. R. Saliva as a diagnostic fluid. *Oral Dis, 8:* 69–76, 2002.

16. Lawrence, H. P. Salivary markers of systemic disease: noninvasive diagnosis of disease and monitoring of general health. *J Can Dent Assoc, 68:* 170–174, 2002.

17. Kaufman, E. and Lamster, I. B. The diagnostic applications of saliva–a review. *Crit Rev Oral Biol Med, 13:* 197–212, 2002.

18. Streckfus, C. F., Bigler, L. R., and Zwick, M. The use of surface-enhanced laser desorption/ionization time-of-flight mass spectrometry to detect putative breast cancer markers in saliva: A feasibility study. *J Oral Pathol Med, 35:* 292–300, 2006.

19. Thompson, E. J. Quality versus quantity: Which is better for cerebrospinal fluid IgG? *Clin Chem*, 50: 1721–1722, 2004.
20. Romeo, M. J., Espina, V., Lowenthal, M., Espina, B. H., Petricoin, E. F., 3rd, and Liotta, L. A. CSF proteome: A protein repository for potential biomarker identification. *Expert Rev Proteomics*, 2: 57–70, 2005.
21. Guerreiro, N., Gomez-Mancilla, B., and Charmont, S. Optimization and evaluation of surface-enhanced laser-desorption/ionization time-of-flight mass spectrometry for protein profiling of cerebrospinal fluid. *Proteome Sci*, 4: 7, 2006.
22. Richter, R., Schulz-Knappe, P., Schrader, M., Standker, L., Jurgens, M., Tammen, H., and Forssmann, W. G. Composition of the peptide fraction in human blood plasma: database of circulating human peptides. *J Chromatogr B Biomed Sci Appl*, 726: 25–35, 1999.
23. Kong, F., Nicole White, C., Xiao, X., Feng, Y., Xu, C., He, D., Zhang, Z., and Yu, Y. Using proteomic approaches to identify new biomarkers for detection and monitoring of ovarian cancer. *Gynecol Oncol*, 100: 247–253, 2006.
24. Ye, B., Cramer, D. W., Skates, S. J., Gygi, S. P., Pratomo, V., Fu, L., Horick, N. K., Licklider, L. J., Schorge, J. O., Berkowitz, R. S., and Mok, S. C. Haptoglobin-alpha subunit as potential serum biomarker in ovarian cancer: identification and characterization using proteomic profiling and mass spectrometry. *Clin Cancer Res*, 9: 2904–2911, 2003.
25. Voshol, H., Brendlen, N., Muller, D., Inverardi, B., Augustin, A., Pally, C., Wieczorek, G., Morris, R. E., Raulf, F., and van Oostrum, J. Evaluation of biomarker discovery approaches to detect protein biomarkers of acute renal allograft rejection. *J Proteome Res*, 4: 1192–1199, 2005.
26. de Seny, D., Fillet, M., Meuwis, M. A., Geurts, P., Lutteri, L., Ribbens, C., Bours, V., Wehenkel, L., Piette, J., Malaise, M., and Merville, M. P. Discovery of new rheumatoid arthritis biomarkers using the surface-enhanced laser desorption/ionization time-of-flight mass spectrometry ProteinChip approach. *Arthritis Rheum*, 52: 3801–3812, 2005.
27. Yang, S. Y., Xiao, X. Y., Zhang, W. G., Zhang, L. J., Zhang, W., Zhou, B., Chen, G., and He, D. C. Application of serum SELDI proteomic patterns in diagnosis of lung cancer. *BMC Cancer*, 5: 83, 2005.
28. Zhang, H., Kong, B., Qu, X., Jia, L., Deng, B., and Yang, Q. Biomarker discovery for ovarian cancer using SELDI-TOF-MS. *Gynecol Oncol*, 102: 61–66, 2006.
29. Murphy, V. E., Johnson, R. F., Wang, Y. C., Akinsanya, K., Gibson, P. G., Smith, R., and Clifton, V. L. The effect of maternal asthma on placental and cord blood protein profiles. *J Soc Gynecol Investig*, 12: 349–355, 2005.
30. Zhang, Y. F., Wu, D. L., Guan, M., Liu, W. W., Wu, Z., Chen, Y. M., Zhang, W. Z., and Lu, Y. Tree analysis of mass spectral urine profiles discriminates transitional cell carcinoma of the bladder from noncancer patient. *Clin Biochem*, 37: 772–779, 2004.
31. Clarke, W., Silverman, B. C., Zhang, Z., Chan, D. W., Klein, A. S., and Molmenti, E. P. Characterization of renal allograft rejection by urinary proteomic analysis. *Ann Surg*, 237: 660–664; discussion 664–665, 2003.

32. Schaub, S., Wilkins, J., Weiler, T., Sangster, K., Rush, D., and Nickerson, P. Urine protein profiling with surface-enhanced laser-desorption/ionization time-of-flight mass spectrometry. *Kidney Int, 65:* 323–332, 2004.

33. Kemna, E., Tjalsma, H., Laarakkers, C., Nemeth, E., Willems, H., and Swinkels, D. Novel urine hepcidin assay by mass spectrometry. *Blood, 106:* 3268–3270, 2005.

34. Dare, T. O., Davies, H. A., Turton, J. A., Lomas, L., Williams, T. C., and York, M. J. Application of surface-enhanced laser desorption/ionization technology to the detection and identification of urinary parvalbumin-alpha: a biomarker of compound-induced skeletal muscle toxicity in the rat. *Electrophoresis, 23:* 3241–3251, 2002.

35. Ye, B., Skates, S., Mok, S. C., Horick, N. K., Rosenberg, H. F., Vitonis, A., Edwards, D., Sluss, P., Han, W. K., Berkowitz, R. S., and Cramer, D. W. Proteomic-based discovery and characterization of glycosylated eosinophil-derived neurotoxin and COOH-terminal osteopontin fragments for ovarian cancer in urine. *Clin Cancer Res, 12:* 432–441, 2006.

# 4

# A Novel Approach Using MALDI-TOF/TOF Mass Spectrometry and Prestructured Sample Supports (AnchorChip Technology) for Proteomic Profiling and Protein Identification

Sau-Mei Leung and Rebecca L. Pitts

## Summary

Mass spectrometry (MS)-based proteomic profiling and protein identification has become a powerful tool for the discovery of new disease biomarkers. Among the MS platforms, matrix-assisted laser desorption/ionization time-of-flight/time-of-flight (MALDI-TOF/TOF) MS offers high sample throughput and the flexibility to couple with different off-line sample fractionation techniques. Here, we present a strategy using MALDI-TOF/TOF MS to analyze fractionated human serum samples for proteomic profiling and then identify serum peptides from these proteomic profiles. We achieve the profiling analyses by using different functionalized magnetic beads to enrich specific subsets of serum proteins/peptides based on their absorption to these beads. This step is followed by elution, transfer onto prestructured sample supports (AnchorChip™ targets), and analysis in a MALDI-TOF/TOF mass spectrometer. Selected serum peptides are then analyzed in the tandem MS (TOF/TOF) mode to generate fragment ions for determination of their amino acid sequences. We have demonstrated that using this approach, proteomic profiling and protein identification can be done in a single MS instrument.

**Key Words:** MALDI; TOF/TOF; profiling; biomarkers; AnchorChip; tandem MS; magnetic beads; clinical proteomics; serum.

## 1. Introduction

Matrix-assisted laser desorption/ionization time-of-flight (MALDI-TOF) and surface-enhanced laser desorption/ionization (SELDI)-TOF mass spectrometry (MS) are increasingly being used to simultaneously profile large numbers of

From: *Methods in Molecular Biology, vol. 441: Tissue Proteomics: Pathways, Biomarkers, and Drug Discovery*
Edited by: B.C.-S. Liu © Humana Press, Totowa, NJ

proteins and peptides for biomarker discovery *(1–9).* Sequencing of profiled peptides by MALDI time-of-flight/time-of-flight (TOF/TOF) MS provides an additional dimension of information content to these proteomic profiles, which is crucial in the biomarker discovery process. Like other MALDI-TOF MS applications, sample preparation is crucial to generate useful and reproducible proteomic profiles and the subsequent tandem MS spectrum for protein identification. In this chapter, we describe a sample preparation method that uses functionalized magnetic beads combined with prestructured MALDI sample supports (AnchorChip technology) to reduce the complexity of serum while remaining easily amenable to automated high-throughput applications *(8).* Samples are fractionated using a variety of magnetic beads containing different surface chemistries [e.g., hydrophobic interaction, cation exchange, and immobilized metal affinity chromatography (IMAC)] to selectively enrich a subset of proteins/peptides from serum based on their physical, chemical, or biological properties. Then, the bound proteins/peptides are eluted from the magnetic beads, mixed with matrix, and spotted onto the AnchorChip target. The AnchorChip target is equipped with hydrophilic spots (anchors) surrounded by a hydrophobic coating to keep the sample solution from spreading. Typically 0.5–2 µL of the matrix/analyte sample is deposited onto the anchor. As the sample droplet shrinks during solvent evaporation, it centers itself onto the anchor. AnchorChip technology has already been shown to improve sensitivity compared with conventional stainless steel supports and provides more uniform MALDI sample preparation, which aids automatic data acquisition *(10–12).* Next, we use TOF/TOF as mass analyzer for the mass spectrometer because it offers flexibility for both proteomic profiling and protein identification applications in a single instrument that has good sensitivity, mass accuracy, and resolution. In addition, the instrument is relatively easy to operate in both manual and automatic data acquisition modes. Our MALDI-TOF/TOF mass spectrometer can be operated in linear mode, reflector mode, or tandem MS (TOF/TOF) mode. In the linear mode, the ions travel down to the linear flight path and are detected by the linear detector. In a reflector mode MALDI instrument, an ion mirror (reflector) is placed at its end, which reflects the ions back to a reflector detector. One of the major advantages of the reflector mode is that it permits higher mass accuracy and resolution measurement. In the tandem MS mode, fragment ions are generated from the precursor ions and this permits the determination of amino acid sequences of the precursor ions *(13,14).* Because the first goal of proteomic profiling is to maximize the number of peaks being detected, we tune the linear mode method in the MALDI-TOF/TOF MS for maximum sensitivity. Once a good-quality linear proteomic profile is generated, we switch to a reflector mode in the instrument to accurately measure the isotopically resolved masses of the peptides that are then chosen for sequencing in TOF/TOF mode for protein identification. Using this approach, we are able to generate reproducible proteomic profiles and successfully sequence peptides directly from the proteomic profile. Our initial sequencing experiments were carried out using serum samples fractionated by IMAC-Copper (Cu) magnetic beads, and selected peptides were analyzed by using MALDI-TOF/TOF MS. We have identified one peptide at m/z 1465 as a sequence from alpha-fibrinogen (*see* **Fig. 1**). In summary, we find that sample preparation (magnetic beads fractionation coupled with AnchorChip target preparation) and instrument tuning are both critical to generate high-quality data.

## 2. Materials

### 2.1. Preparation of Matrix

1. Matrix: alpha-cyano-4-hydroxycinnamic acid (HCCA, ~1 mg)
2. 100% acetone (500 μL, HPLC grade)
3. 100% ethanol (500 μL, HPLC grade)
4. 1.5 mL microcentrifuge tubes (*see* **Note 1**)

### 2.2. Tandem Fractionation Using IMAC-Copper (IMAC-Cu) and Hydrophobic Interaction Chromatography-C8 (HIC-C8) Magnetic Beads Kits

1. Human serum sample (20 μL, *see* **Note 2**)
2. IMAC-Cu and HIC-C8 magnetic beads kit (1 each, Bruker Daltonics, Billerica, MA, USA, *see* **Note 3**)
3. Magnetic beads separator (8- or 96-wells, Bruker Daltonics)
4. Thin-wall polymerase chain reaction (PCR) 8-well strip tube (*see* **Note 1**)
5. 50% acetonitrile (ACN, HLPC grade) in deionized (18 MΩ) water

### 2.3. Matrix/Analyte Dilution and Spotting Onto AnchorChip Target

1. 0.5-mL Eppendorf tubes for matrix/analyte dilution (*see* **Note 1**)
2. AnchorChip target, 384 spots with 600 μm anchors (Bruker Daltonics, *see* **Note 4**)

### 2.4. Preparation of Calibration Standard

1. Peptide calibration standards (Bruker Daltonics) containing the following calibrants ([M+H]$^+$ monoisotopic mass, Daltons): angiotensin II (1046.5418), angiotensin I (1296.6848), substance P (1347.7354), bombesin (1619.8223), ACTH clip 1–17 (2093.0862), ACTH clip 18–39 (2465.1983), somatostatin 28 (3147.4710).
2. 0.1% trifluoroacetic acid (TFA, sequencing grade) in deionized water (200 μL)

**A.**

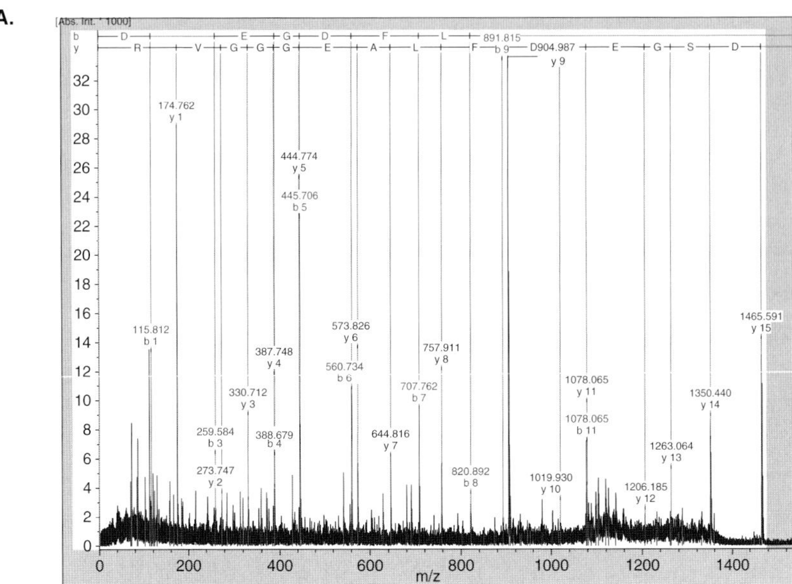

**B.**

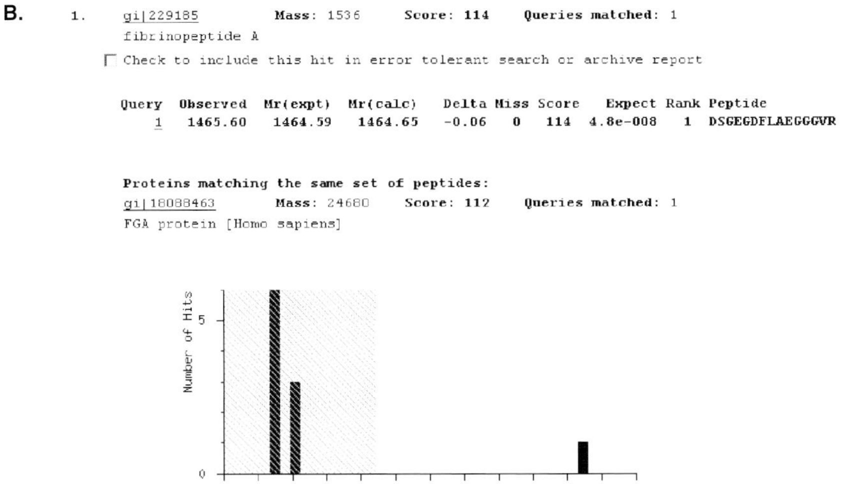

Fig. 1. Identification of m/z 1465 from proteomic profile. Panel A shows the TOF/TOF spectrum of m/z 1465 annotated in BioTools (Bruker Daltonics, Inc.). Panel B shows results from the MASCOT search engine.

## 2.5. General Supplies

1. Pipette tips (*see* **Note 1**)
2. Aspiration station
3. Ice bucket for keeping the serum sample and eluted samples at 4 °C
4. Timer
5. Powder-free gloves

# 3. Methods

All procedures are carried out at room temperature (*see* **Note 5**).

## 3.1. Preparation of Matrix

1. HCCA (0.67 mg/mL in ethanol : acetone, 2:1). Dissolve approximately 1 mg HCCA in 500 μL acetone to make a 2-mg/mL stock solution. The matrix usually takes 5–10 min to completely dissolve at room temperature. We usually start the magnetic beads fractionation while waiting for the matrix to dissolve. Then add one part of above stock solution to two parts ethanol (e.g. 100 μL stock solution to 200 μL ethanol). Prepare the matrix solution daily.

## 3.2. Tandem Fractionation Using IMAC-Cu and HIC-C8 Magnetic Beads Kits

### 3.2.1. Preparation of IMAC-Cu Magnetic Beads

1. Add 50 μL IMAC-Cu-binding solution to a thin-wall PCR tube.
2. Mix the magnetic beads thoroughly to get a homogenous solution by shaking the tube. Pipette 5 μL (100 mg/mL) of magnetic beads into the above thin-wall PCR tube. Repeat shaking between pipetting steps when necessary.
3. Place the tube in a magnetic bead separator and wash the beads by moving the tube back and forth between adjacent wells 10 times. (Note the movement of the magnetic beads in the tube.)
4. Keep the tube in the magnetic bead separator for 20 s without moving to allow the beads collected at one side of the tube wall (i.e., to separate the beads from the supernatant).
5. Discard the supernatant.
6. Repeat **steps 3–5** two additional times.
7. Finally, resuspend the magnetic beads in 20 μL IMAC-Cu-binding solution.

### 3.2.2. Preparation of HIC-C8 Magnetic Beads

1. Add 10 μL HIC-C8-binding solution to a thin-wall PCR tube.
2. Mix the magnetic beads thoroughly to get a homogenous solution by shaking the tube. Pipette 5 μL (50 mg/mL) of magnetic beads to the above thin-wall PCR tube. Repeat shaking between pipetting steps when necessary.

### 3.2.3. Sample Fractionation

1. Add 5 μL human serum sample to IMAC-Cu magnetic beads and mix carefully by pipetting up and down five times.
2. Keep at room temperature for 5 min.
3. Place the tube in a magnetic beads separator and wait for 20 s to allow the beads to separate from the supernatant.
4. Transfer all the supernatant (~25 μL of IMAC-Cu unbound solution) to HIC-C8 magnetic beads. Mix by pipetting up and down five times. Wait for 1 min.
5. Place the tube in a magnetic beads separator and wait for 20 s to allow the beads to separate from the supernatant. Discard the supernatant.

### 3.2.4. Wash the Unbound Analytes from Magnetic Beads

1. Add 100 μL IMAC-Cu or HIC-C8 wash solution to the respective tube and place the tubes in a magnetic beads separator. Again, wash the beads by moving tubes back and forth between adjacent wells 10 times. (Note the movement of the magnetic beads in the tube.)
2. Keep the tube in the magnetic bead separator for 20 s without moving to allow the beads collected at one side of the tube wall (i.e., to allow the beads to separate from the supernatant).
3. Discard the wash solution.
4. Repeat **steps 1–3** two additional times.

### 3.2.5. Elute the Bound Analytes

1. Remove the tubes from the magnetic beads separator.
2. Add 10 μL IMAC-Cu elution solution to the IMAC-Cu tube and mix thoroughly by pipetting up and down five times.
3. Wait for 5 min.
4. Add 5 μL 50% ACN to the HIC-C8 tube and mix thoroughly by pipetting up and down five times.
5. Wait for 1 min.
6. Place the tubes in a magnetic beads separator and wait for about 20 s to separate the beads from the elution solution at the wall of the tubes.
7. Transfer the eluted samples into a fresh tube and cap. Keep the tubes on ice to prevent evaporation.

### 3.3. Matrix/Analyte Dilution and Spotting onto AnchorChip Target

1. Mix 1 μL of eluted sample and 9 μL matrix solution (from **Subheading 3.1.**) and spot 1 μL onto AnchorChip target. Continue spotting the rest of the mixture onto the target (~8 spots). Let the samples dry at room temperature (*see* **Note 6**).

### *3.4. Preparation of Calibration Standards*

1. Dissolve the dried peptide calibration standards by adding 125 µL 0.1% TFA to the tube (final concentration will be 4 pmol/µL for each peptide). Mix 1 µL of peptide standard and 9 µL matrix solution and spot 1 µL onto the "calibration spots" of the AnchorChip target. The final concentration of peptide standard on the target is 400 fmol of each peptide.

### *3.5. Data Acquisition Using MALDI-TOF/TOF Mass Spectrometer (ultraflex TOF/TOF, Bruker Daltonics)*

#### *3.5.1. Proteomic Profiling*

1. Instrument parameters: Proteomic profiles are acquired in linear positive ion mode using automated data collection with fixed laser energy operating at 25 Hz with a 337-nm nitrogen laser (*see* **Note 7**). Ions are accelerated at 25 kV with 200 ns of pulsed ion extraction delay with the extraction voltage at 23 kV. The lens voltage is 8 kV, and the linear detector voltage is 1.56 kV. The matrix deflection mass is 800 kDa, and the mass range for acquisition is mass-to-charge ratio (m/z) 1000–20,000 (*see* **Note 8**).
2. Instrument calibration: The instrument is externally calibrated using a linear fit equation with a mixture of six peptide standards (Bruker Daltonics) before data acquisition of the samples (*see* **Subheading 2.4.**).
3. Automatic data acquisition parameters: To increase detection sensitivity, excess matrix is removed from the sample spots with 10 shots at higher laser power (matrix blaster shots approximately 10–20% higher laser energy) before acquisition of spectra with 300 shots (30 shots × 10 different locations per spot) per spectrum. The matrix blaster shots are not added to the final spectrum. It is very important to adjust the laser power to obtain a good quality of spectrum (*see* **Fig. 2**, and **Notes 9 and 10**).

#### *3.5.2. Protein Identification*

1. Instrument parameter settings: For the reflector positive ion mode, ions are accelerated at 25 kV with 50 ns of pulsed ion extraction delay with the extraction voltage at 21.75 kV. The lens voltage is 9.5 kV and the reflector voltages 1 and 2 are 26.3 and 14.1 kV, respectively. The reflector detector voltage is 1.625 kV, and the laser is operated at 50 Hz repetition rate. The matrix deflection mass is 500 kDa and the mass range for acquisition is m/z 580 to 4000. We used the default LIFT method *(13)* for the tandem MS (TOF/TOF mode) analysis.
2. Instrument calibration: The instrument is calibrated using close external calibration using a quadratic fit equation with a mixture of six peptide standards (Bruker Daltonics) before data acquisition of the samples. Adjust the laser to get some good spectra for the standards (*see* **Note 11**) with an average mass accuracy less than 10 ppm.

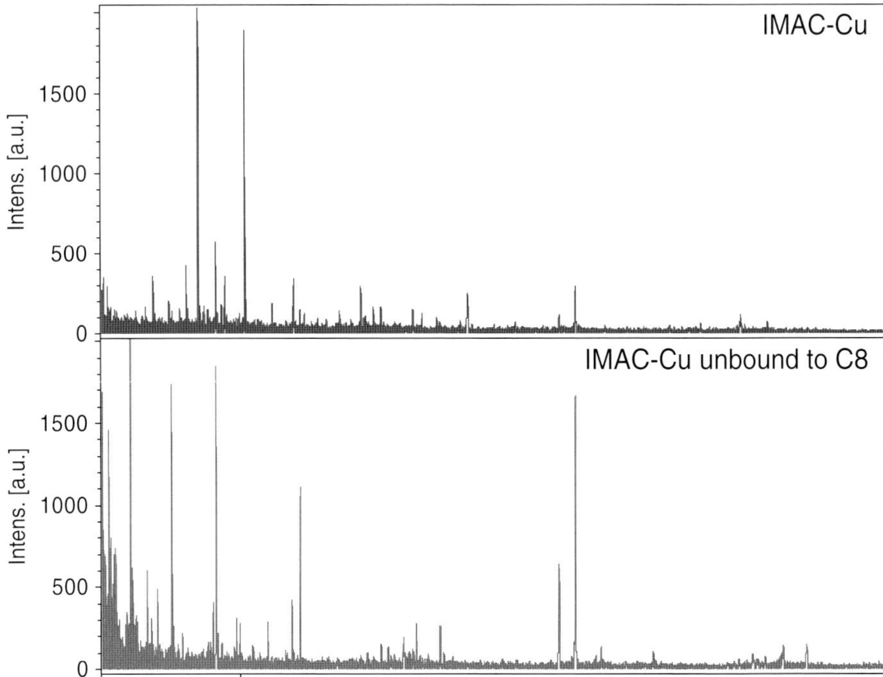

Fig. 2. Mass spectra (m/z 1000–10,000) acquired by MALDI-TOF/TOF MS in linear mode using serum sample fractionated by IMAC-Cu and hydrophobic C8 magnetic beads. The serum sample was fractionated using IMAC-Cu magnetic beads first and then the unbound solution was transferred to C8 magnetic beads for additional fractionation (tandem chromatography). Different profiles are generated using this tandem chromatography method.

3. After acquiring the proteomic profiles in linear mode, peptides below m/z 4000 with good peak intensity ($\sim$1000–2000 arbitrary intensity unit, *see* **Fig. 1**) are selected for fragmentation analysis using the TOF/TOF mode for protein identification.

4. The first step in a tandem MS experiment, in this case, a TOF/TOF experiment, is to determine the exact mass of the selected peptides (precursor ion). The monoisotopic mass of the precursor ion is measured in positive, reflector mode using close external calibration (*see* **Note 11**).

5. Once the exact mass is determined for the precursor ion, the instrument is switched to tandem MS mode (TOF/TOF mode) and fragment ions are generated using laser-induced dissociation (LID) and/or high energy collision-induced dissociation (CID) *(13)* (*see* **Note 12**).

6. The resulting tandem MS spectrum is labeled in FlexAnalysis and exported to BioTools (annotation software for the MS/MS spectrum). The peak list generated

by the BioTools software is automatically downloaded to the MASCOT software [Matrix Science Ltd., London, UK *(15)*] for database searching.

7. Database search parameters for the TOF/TOF data using the Mascot search engine: *Homo sapiens* for taxonomy, NCBIr for database, no enzyme, no modifications, mass tolerance of the precursor ion at ± 0.2 kDa and the fragment ions at ± 0.8 kDa.

## 4. Notes

1. It is very important to "quality control" the plasticware that comes in contact with the samples, buffers, and solvents because lubricants and plasticizers added to the plastic resin during the manufacturing process may leak out and possibly affect the quality of the spectrum *(16)*. We use colorless polypropylene pipette tips, microcentrifuge tubes, and thin-wall PCR strip tubes from Eppendorf. For storage of buffers and solvents, we use Nalgene Teflon bottles with Telfon caps. We also use Corning PYREX bottles with Teflon-lined caps to store buffers.

2. The serum is aliquoted 20 µL per tube, stored at –70 °C and used only once after thawing. Several articles have discussed the importance of standardization of preanalytical factors such as patient preparation and sample collection methods for MS-based proteomics profiling *(17–20)*.

3. The magnetic beads kit is supplied with a standard protocol and binding, washing, and elution solutions.

4. The AnchorChip target is reusable. We clean the target using the following procedure:

   i. Carefully wipe the target with a suitable solvent (e.g., acetone for HCCA matrix) using a soft tissue (e.g., Kimwipe) to remove the majority of analyte/matrix crystals.
   ii. Sonicate the target in 50% methanol for about 5–10 min.
   iii. Rinse the target in methanol.
   iv. Rinse the target in deionized water.r

To make sure the target is clean, we spot matrix onto the target and then collect data in the mass spectrometer to see whether any sample contamination appears.

5. We keep the laboratory temperature at approximately at 21 °C (70 °F). It may also be helpful if you control the humidity in the laboratory because the rate of co-crystallization of the matrix/analyte may be affected by the humidity that will affect the quality of the spectrum.

6. We found that target preparation choices such as type and concentration of matrix, the ratio of analyte to matrix, on-target washing, and recrystallization procedures can dramatically affect the quality of resulting MALDI spectra. We usually try two different analyte to matrix dilutions: 1–9 µL and 1–4 µL (analyte to matrix) to ensure good-quality spectra. We tune the linear mode method in

the MALDI-TOF/TOF MS for maximum sensitivity to maximize the number of peaks being detected with good peak intensity (~1000–2000 arbitrary intensity unit, *see* **Fig. 1**). If the resulting spectrum is not good, we try washing and recrystallizing the sample on the AnchorChip target. On-target washing is a simple procedure that will remove water soluble contaminants. HCCA crystals with incorporated analyte will not dissolve in the wash solution. The following is the on-target washing procedure:

1. Add 5–10 μL wash solution (0.1% TFA) to the sample spot
2. Remove wash solution after a few seconds without touching the crystals
3. Let the residual liquid evaporate.

You can also recrystallize the analyte/matrix crystal after on-target washing. Add 0.5–1 μL ethanol : acetone : 0.1% TFA (6:3:1) to the sample spot and let it air dry.

7. We find that the higher the laser repetition rate, the lower the intensity of the resulting spectrum for the linear mode-profiling experiment. We usually use between 12.5 and 25 Hz for data collection.
8. To get the best spectra with respect to mass accuracy, mass resolution and sensitivity (peak intensity), the mass spectrometer can be operated in different modes and experimental settings. The optimal settings depend on the nature of the sample and the goal of the experiment; finding these optimal conditions may be a matter of trial and error. We focus the profiling experiment initially in the peptide mass range because we want to take advantage of the high resolution, mass accuracy, and sensitivity of the instrument in this mass range. We usually start with the default settings from the instrument installation and then tune the instrument for maximum sensitivity for profiling experiments and maximum mass resolution and accuracy for protein identifications. Parameters important to MALDI-TOF instrument performance that must be optimized for any serum proteomic profiling study include the laser repetition rate, laser power, matrix blaster shots, number of shots, pulsed ion extraction delay time, and the plate potentials used to push ions into the flight tube. Once the linear method is tuned, usually the only remaining parameter to optimize for each experiment is the laser intensity. You should also periodically check your instrument performance. You can use the QC tests for mass resolution, sensitivity, and mass accuracy used during the instrument installation as a guideline. It is also a good idea to record the number of the laser shots that have been fired (since the laser has a shelf-life) and the vacuum pressure. Keep a logbook next to the instrument so that user can write down problems and how they resolve them as well as services that have been done on the instrument. Good record keeping is very important to ensure the performance of the instrument, especially when the instrument is being used by multiple researchers.
9. A reproducibility study is the key to successful proteomic pattern analysis using MS *(21,22)*. We perform reproducibility studies using a single sample to establish

coefficient of variance (CV) levels. We divide a single serum sample into three separate tubes and generate a proteomic profile for each tube. Then, the CV is calculated based on 10 randomly selected peaks from m/z 1000–10,000. Using manual preparation, we usually generate an average peak area CV below 26% *(8)*. We believe that success in proteomic profiling requires a good understanding of:

   i. Sample components and complexity
   ii. Surface chemistry of the magnetic beads
   iii. Bind/wash/elute conditions
   iv. MALDI target preparation (choice of matrix, solvents, matrix concentration, preparation methods, etc.)
   v. Data acquisition (optimization of instrument settings).

Different samples may require different preparation techniques. We take the following steps when we encounter an unknown sample for the first time or try to trouble shoot with problematic samples:

   a. We usually start with the standard protocol from three magnetic beads kits (C8, IMAC-Cu and Weak Cation eXchange, WCX) and then try a 1:9 or 1:4 analyte : matrix preparation. We call this a scouting experiment and usually find one of the above combinations will generate a good-quality spectrum.
   b. We then try on-target washing and recrystallization if none of the scouting experiment combinations generate good spectra.
   c. If the on-target washing and recrystallization fail to improve the spectrum quality, we then vary the matrix to analyte ratios (e.g., 1:1, 1:2, or 1:20) to investigate whether the eluted sample is too dilute or too concentrated. We have also tried spotting 2–3 µL of matrix/analyte solution onto the target.
   d. We have also tried using matrix concentration (e.g., HCCA) at 1.2 mg/mL instead of 0.6 mg/mL.

10. MALDI-TOF/TOF MS is a very sensitive instrument and is able to measure attomoles to femtomoles of purified material. However, the signal achieved is both a function of concentration and purity. Ion suppression can be described as when two proteins or peptides are present as a mixture in MALDI analysis, one may predominate over the other in the mass spectrum. In some cases, a protein or peptide can be observed if it is loaded as a pure sample, but if it is loaded at the same amount in a mixture with other proteins or peptides, its intensity may be suppressed. In general, MALDI is more tolerable to biological buffers *(23)*, but nonvolatile solution such as glycerol or DMSO must be avoided. If a detergent is unavoidable, 0.1% octyl-beta-D-glucopyranoside can be used *(24–27)*. For MALDI, the detection of peptides has been shown to be strongly dependent on amino acid composition. The presence of the basic residues (e.g., arginine) greatly increases ionization efficiency and hence, peak signal *(28,29)*.

11. We used the instrument installation mass resolution and accuracy specifications as a guideline to define a good spectrum for protein identification. A good

spectrum in the reflector TOF mode has an average mass accuracy less than 10 ppm using close external calibration and mass resolution for a mixture of peptide standards greater than 10,000 for angiotensin II (1046.5418), 20,000 for ACTH clip 18–39 (2465.1983), and somatostatin 28 (3147.4710).

12. Because only a fraction of the eluted samples was used for analysis, this magnetic bead-based sample fractionation method greatly reduces the sample volume required for generating proteomic profiling and preliminary sequence information. We found that if the peak intensity is small, we try to accumulate shots from multiple spots for a single tandem MS spectrum or we try other enrichment methods to obtain enough material for tandem MS analysis. For example, using scale-up of the magnetic beads fractionation combined with or without sodium dodecyl sulfate–SDS–polyacrylamide gel electrophoresis (PAGE) may be able to enrich the particular peptides of interest. Other approaches such as LC-MALDI-TOF/TOF MS can also be used to further fractionate or concentrate peptides for tandem MS analysis. However, if there is enough material from the magnetic beads eluted sample, using MALDI-TOF/TOF MS is a very rapid way to obtain sequence information (peptides less than m/z 4000) from proteomic profiles. If the proteins/peptides of interest are greater than m/z 4000, we suggest using different separation techniques (magnetic beads fractionation, gel electrophoresis, and/or chromatography) to purify the proteins/peptides of interest and then perform peptide mass fingerprints (PMF) and sequencing of the digested peptides using MALDI-TOF/TOF MS for protein identification.

## References

1. Adam, B. L., Qu, Y., Davis, J. W., Ward, M. D., Clements, M. A., Cazares, L. H., Semmes, O. J., Schellhammer, P. F., Yasui, Y., Feng, Z., and Wright, G. L., Jr. (2002) Serum protein fingerprinting coupled with a pattern-matching algorithm distinguishes prostate cancer from benign prostate hyperplasia and healthy men. *Cancer Res.* **62**, 3609–3614.

2. Petricoin, E. F., Ardekani, A. M., Hitt, B. A., Levine, P. J., Fusaro, V. A., Steinberg, S. M., Mills, G. B., Simone, C., Fishman, D. A., Kohn, E. C., and Liotta, L. A. (2002) Use of proteomic patterns in serum to identify ovarian cancer. *Lancet* **359**, 572–577.

3. Ye, B., Cramer, D. W., Skates, S. J., Gygi, S. P., Pratomo, V., Fu, L., Horick, N. K., Licklider, L. J., Schorge, J. O., Berkowitz, R. S., and Mok, S. C. (2003) Haptoglobin-alpha subunit as potential serum biomarker in ovarian cancer: identification and characterization using proteomic profiling and mass spectrometry. *Clin. Cancer Res.* **9**, 2904–2911.

4. Zhang, Z., Bast, R. C., Jr., Yu, Y., Li, J., Sokoll, L. J., Rai, A. J., Rosenzweig, J. M., Cameron, B., Wang, Y. Y., Meng, X. Y., Berchuck, A., Van Haaften-Day, C., Hacker, N. F., de Bruijn, H. W., van der Zee, A. G., Jacobs, I. J., Fung, E. T., and Chan, D. W. (2004) Three biomarkers identified from serum proteomic analysis for the detection of early stage ovarian cancer. *Cancer Res.* **64**, 5882–5890.

5. Xu, X. Q., Leow, C. K., Lu, X., Zhang, X., Liu, J. S., Wong, W. H., Asperger, A., Deininger, S., and Eastwood Leung, H. C. (2004) Molecular classification of liver cirrhosis in a rat model by proteomics and bioinformatics. *Proteomics* **4**, 3235–3245.

6. Villanueva, J., Philip, J., Entenberg, D., Chaparro, C. A., Tanwar, M. K., Holland, E. C., and Tempst, P. (2004) Serum peptide profiling by magnetic particle-assisted, automated sample processing and MALDI-TOF mass spectrometry. *Anal. Chem.* **76**, 1560–1570.

7. Pusch, W., Flocco, M. T., Leung, S.-M., Thiele, H., and Kostrzewa, M. (2003) Mass spectrometry-based clinical proteomics. *Pharmacogenomics* **4**, 1–14.

8. Zhang, X. Y., Leung, S.-M., Morris, C. R., and Shigenaga, M. K. (2004) Evaluation of a novel, integrated approach using functionalized magnetic beads, bench-top MALDI-TOF MS with prestructured sample supports and pattern recognition software for profiling potential biomarkers in human plasma. *J. Biomol. Tech.* **15**, 167–175.

9. Koomen, J. M., Shih, L. N., Coombes, K. R., Li, D., Xiao, L. C., Fidler, I. J., Abbruzzese, J. L., and Kobayashi, R. (2005) Plasma protein profiling for diagnosis of pancreatic cancer reveals the presence of host response proteins. *Clin. Cancer Res.* **11**, 1110–1118.

10. Schuerenberg, M., Luebbert, C., Eickhoff, H., Kalkum, M., Lehrach, H., and Nordhoff, E. (2000) Prestructured MALDI-MS sample supports. *Anal. Chem.* **72**, 3436–3442.

11. Gobom, J., Schuerenberg, M., Mueller, M., Theiss, D., Lehrach, H., and Nordhoff, E. (2001) Alpha-Cyano-4-HCCA affinity sample preparation. A protocol for MALDI-MS peptide analysis in proteomics. *Anal. Chem.* **73**, 434–438.

12. Nordhoff, E., Schuerenberg, M., Thiele, G.; Lubbert, C., Kloeppel, K.-D., Theiss, D., Lehrach, H., and Gobom, J. (2003) Sample preparation protocols for MALDI-MS of peptides and oligonucleotides using prestructured sample supports. *Int. J. Mass. Spectrom.* **226**, 163–180.

13. Suckau, D., Resemann, A., Schuerenberg, M., Hufnagel, P., Franzen, J., and Holle, A. (2003) A novel MALDI LIFT-TOF/TOF mass spectrometer for proteomics. *Anal. Bioanal. Chem.* **376**, 952–965.

14. Macht, M., Asperger, A., and Deininger, S. O. (2004) Comparison of laser-induced dissociation and high-energy collision-induced dissociation using matrix-assisted laser desorption/ionization tandem time-of-flight (MALDI-TOF/TOF) for peptide and protein identification. *Rapid Commun. Mass Spectrom.* **18**, 2093–2105.

15. http://www.matrixscience.com/

16. Drake, S. K., Bowen, R. A., Remaley, A. T., and Hortin, G. L. (2004) Potential interferences from blood collection tubes in mass spectrometric analyses of serum polypeptides. *Clin. Chem.* **50**, 2398–2401.

17. Schaub, S., Wilkins, J., Weiler, T., Sangster, K., Rush, D., and Nickerson, P. (2004) Urine protein profiling with surface-enhanced laser-desorption/ionization time-of-flight mass spectrometry. *Kidney Int.* **65**, 323–332.

18. Drake, R. R., Cazares, L. H., Corica, A., Malik, G., Schwegler, E. E., Libby, A. E., Wright, G. L., Jr., Adam, B.-L., and Semmes, O. J. (2004) Quality control,

preparation and protein stability issues for blood serum and plasma used in biomarker discovery and proteomic profiling assay. *Bioprocessing J.* **3**, 45–50.

19. Hulmes, J. D., Bethea, D., Ho, K., Huang, S.-P., Ricci, D. L., Opiteck, G. J., and Hefta, S. A. (2004) An investigation of plasma collection, stabilization, and storage procedures for proteomic analysis of clinical samples. *Clin. Proteomics J.* **1**, 17–31.
20. Hortin G. L. (2005) Can mass spectrometric protein profiling meet desired standards of clinical laboratory practice? *Clin. Chem.* **51**, 3–5.
21. Semmes, O. J., Feng, Z., Adam, B. L., Banez, L. L., Bigbee, W. L., Campos, D., Cazares, L. H., Chan, D. W., Grizzle, W. E., Izbicka, E., Kagan, J., Malik, G., McLerran, D., Moul, J. W., Partin, A., Prasanna, P., Rosenzweig, J., Sokoll, L. J., Srivastava, S., Srivastava, S., Thompson, I., Welsh, M. J., White, N., Winget, M., Yasui, Y., Zhang, Z., and Zhu, L. (2005) Evaluation of serum protein profiling by surface-enhanced laser desorption/ionization time-of-flight mass spectrometry for the detection of prostate cancer: I. Assessment of platform reproducibility. *Clin. Chem.* **51**, 102–112.
22. Dekker, L. J., Dalebout, J. C., Siccama, I., Jenster, G., Sillevis Smitt, P. A., and Luider, T. M. (2005) A new method to analyze matrix-assisted laser desorption/ionization time-of-flight peptide profiling mass spectra. *Rapid Commun. Mass Spectrom.* **19**, 865–870.
23. List for concentration tolerances of buffers components for MALDI analysis using dried droplet method (http://w3.ouhsc.edu/biochem/sampleprep.htm).
24. Cohen, S. L. and Chait, B. T. (1996) Influence of matrix solution conditions on the MALDI-MS analysis of peptides and proteins. *Anal. Chem.* **68**, 31–7.
25. http://prowl.rockefeller.edu/contam/deterg.htm
26. Zhang, N. and Li, L. (2004) Effects of common surfactants on protein digestion and matrix-assisted laser desorption/ionization mass spectrometric analysis of the digested peptides using two-layer sample preparation. *Rapid Commun. Mass Spectrom.* **18**, 889–896.
27. Zhu, Y. F., Lee, K. L., Tang, K., Allman, S. L., Taranenko, N. I., and Chen, C. H. (1995) Revisit of MALDI for small proteins. *Rapid Commun. Mass Spectrom.* **9**, 1315–1320.
28. Krause, E., Wenschuh, H., and Jungblut, P. R. (1999) The dominance of arginine-containing peptides in MALDI-derived tryptic mass fingerprints of proteins. *Anal. Chem.* **71**, 4160–4165.
29. Baumgart, S., Lindner, Y., Kuhne, R., Oberemm, A., Wenschuh, H., and Krause, E. (2004) The contributions of specific amino acid side chains to signal intensities of peptides in matrix-assisted laser desorption/ionization mass spectrometry. *Rapid Commun. Mass Spectrom.* **18**, 863–868.

# 5

# Automated Laser Capture Microdissection for Tissue Proteomics

Adrianna S. Rodriguez, Benjamin H. Espina, Virginia Espina, and Lance A. Liotta

## Summary

Laser Capture Microdissection (LCM) is a technique for isolating pure cell populations from a heterogeneous tissue section or cytological preparation through direct visualization of the cells. This technique is applicable to molecular profiling of diseased and disease-free tissue, permitting correlation of cellular molecular signatures with specific cell populations. DNA, RNA, or protein analysis may be performed with the microdissected tissue by any method with adequate sensitivity.

Automated LCM platforms combine graphical user interfaces and annotation software for visualization of the tissue of interest in addition to robotically controlled microdissection. The principal components of LCM technology are (1) visualization of the cells of interest through microscopy, (2) transfer of laser energy to a thermolabile polymer with formation of a polymer—cell composite, and (3) removal of the cells of interest from the heterogeneous tissue section. Automated LCM is compatible with a variety of tissue types, cellular staining methods, and tissue preservation protocols allowing microdissection of fresh or archival specimens in a high-throughput manner. This protocol describes microdissection techniques compatible with downstream proteomic analyses.

**Key Words:** Cancer; DNA; laser capture microdissection; molecular profiling; proteomics; protein; RNA; tissue heterogeneity.

## 1. Introduction

The fields of genomics and proteomics have been enhanced by cellular isolation techniques such as Laser Capture Microdissection (LCM), which enables the isolation and molecular investigation of pure cell populations through direct

From: *Methods in Molecular Biology, vol. 441: Tissue Proteomics: Pathways, Biomarkers, and Drug Discovery*
Edited by: B.C.-S. Liu © Humana Press, Totowa, NJ

visualization of cells. Molecular profiling of a pure cell population, which reflects the cells' in vivo genomic and proteomic state, is essential for correlating molecular signatures in diseased and disease-free cells *(1–8)*. The tissue microenvironment comprises multiple cell types, in a miniature ecosystem, permitting numerous interactions between diseased and disease-free cells. Host interactions, nutritional status of the individual, and immune system function influence the cell's microenvironment. Examples of molecules mediating cell–cell interactions include secreted factors, cell-surface receptors, and adhesion molecules *(4,9,10)*. Designing therapies directed toward these diseased cells, or alternatively toward the surrounding stromal cells, requires a knowledge of the molecular pathways that may be either up-regulated or down-regulated in a particular cell population. Simply preparing a heterogeneous cell lysate is inadequate for deciphering the molecular signatures of individual cell populations.

Cells carrying a constitutive genetic defect may express a variety of proteins depending on the context of the microenvironment. Cancer is an example of a genetically based disease process, but on a functional level, it is a proteomic disease. This genetic alteration is translated into defective or altered proteins that guide the survival of the diseased cell population. The genetic defect of the cancer cell selectively offers a survival advantage for the cell by altering cellular signaling pathways, possibly driving tumor invasion and metastasis *(4,11,12)*. As expected, these pathways are fluid and dynamic, adapting to the changing microenvironment *(13,14)*. Thus, cells, specifically disease cells, are a product of their microenvironment as well as their genetic constitution. It is imperative to know which cell population is contributing to the signal produced in molecular analyses, whether DNA, RNA, or protein is being analyzed. A minor cell population may actually contribute a high percentage of the signal in any given tissue. LCM permits the selection of normal, pre-malignant, and malignant cells, or disease and disease-free cells, as distinct cell populations from the heterogeneous tissue, enabling a more definitive measure of cell signaling proteins and mRNA *(1,2,15–17)*.

The primary components of LCM technology are (1) visualization of the cells of interest through microscopy, (2) transfer of near-infrared laser energy pulses to a thermolabile polymer with formation of a polymer–cell composite, and (3) removal of the polymer from the tissue surface, which shears the embedded cells of interest away from the heterogeneous tissue section *(18,19)*. Extraction buffers applied to the polymer film solubilize the cells, liberating the molecules of interest. The DNA, RNA, or protein from the microdissected cells may be analyzed by any method with appropriate sensitivity *(20–24)*. Protein extracted from microdissected cells may be used for mass spectrometric analysis, applied to reverse phase protein microarrays, or used for western blot analysis *(25,26)*.

LCM, as developed at the National Institutes of Health, originated as a manual, operator-dependent system (PixCell™Molecular Devices, Sunnyvale,

CA, USA) *(18,19)*. The increasing demand for LCM technology among basic science and clinical researchers led to the development of automated versions of LCM, the AutoPix™ and Veritas™ systems (Molecular Devices). All three systems allow for microdissection of cells from any of the following tissue preparations: frozen tissue sections, formalin-fixed paraffin embedded *(FFPE)* tissue sections, cytology smears, and living in vitro cells. LCM systems are compatible with most common histological staining procedures, including fluorescence detection strategies *(27–29)*.

The automated LCM instruments reduce hands-on time by combining standard LCM apparatus with computer-controlled precision. Through the use of robotic stage, optical, filter, camera, and objective movements, the automated systems permit microdissection of cells from up to three slides, and isolation of up to eight different cell types, in one session. The enhanced computer graphics software provides (1) higher resolution for the visualization of distinct cell morphological features, (2) a "road map image" of the entire slide for documentation and orientation (*see* **Fig. 1**), (3) cell and feature detection algorithms, and (4) software annotation tools for selection of cells *(28,29)*.

The Veritas™ system combines automated LCM with UV Laser Cutting. The relatively low-powered near-infrared laser utilized in LCM technology helps preserve the samples' molecular integrity. Thus, it is optimal for capturing small areas or single cells. On the contrary, the Class IV UV cutting laser, which is used to photo volatize cells surrounding a selected area, is optimal for isolating large tissue areas. In addition, the Veritas™ is equipped with a laser ablation feature, which is used to destroy or ablate unwanted material from a sample before or after microdissection (*see* **Fig. 2**) *(29)*.

LCM systems have evolved to accommodate the needs of diverse research fields. Automated LCM enables researchers to isolate specific cells of interest, without contamination from surrounding cells, for genomic and proteomic analysis.

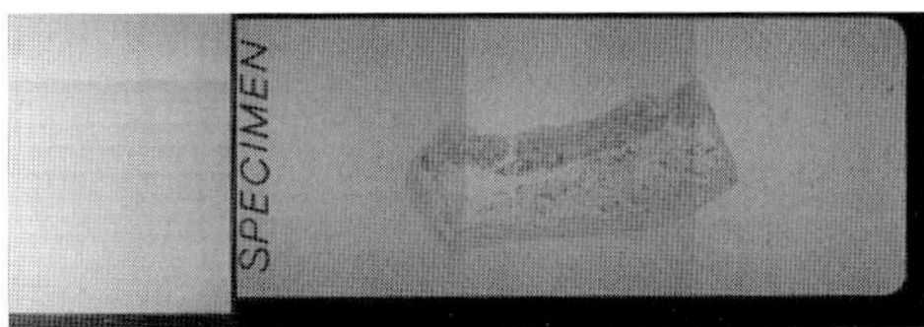


Fig. 1. Roadmap image of the entire slide. Automated Laser Capture Microdissection (LCM) instruments display an image of the entire slide that serves as a reference image for determining the areas of tissue to be microdissected. The tissue shown on the roadmap image is coral. LCM can be performed on a variety of tissue types, including plant, coral, shellfish, and mammalian tissue.

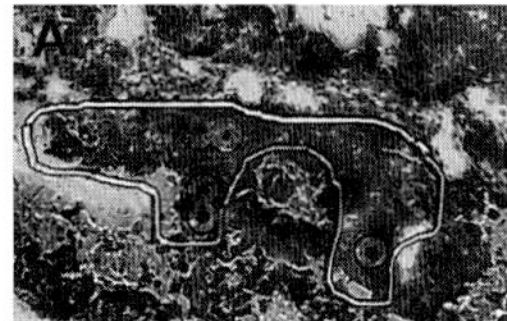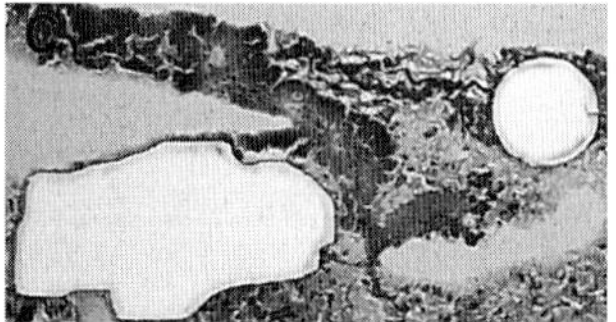

Fig. 2. UV cutting laser tools. (**A**) The UV cutting laser ablates tissue around the perimeter of the desired area. The capture laser is used in conjunction with a Laser Capture Microdissection (LCM) cap to remove the desired tissue section. The LCM cap is bonded, or tacked, to the cut tissue within the perimeter of the cut section, effectively removing the desired cellular area from the tissue section. (**B**) A free-form annotation tool is also available to select any size or shaped region for ablation/cutting with the UV laser. (**C**) Circular regions of defined diameter may be ablated/cut with the UV laser-equipped Veritas™ instrument.

## 2. Materials

### 2.1. Preparation of Tissue Sections or Cytospin Preps

1. Pre-cleaned, uncoated glass microscope slides, 25 × 75 mm (A. Daigger & Co., Wheeling, IL, USA).
2. PEN membrane glass slides (Molecular Devices), for use with the Veritas™ UV cutting laser.
3. Specimen for protein analysis of microdissected tissue: cytospin preps or frozen tissue sections cut at 2–15 μm (5–8 μm is optimal) (*see* **Notes 1** and **2**).
4. Specimen for DNA or RNA analysis of microdissected tissue: cytospin preps, frozen tissue sections, ethanol, or *FFPE* tissue sections cut at 2–15 μm (5–8 μm is optimal) (*see* **Notes 1** and **2**).
5. Cryopreservation solution (OCT) (Sakura Finetek Corp., Torrance, CA, USA).

### 2.2. Hematoxylin and Eosin Staining of Tissue Sections

1. Mayer's Hematoxylin Solution (Sigma Diagnostics, St. Louis, MO, USA). Hematoxylin is an inhalation and contact hazard. Wear gloves when handling.
2. Eosin Y Solution, alcoholic (Sigma Diagnostics). Eosin Y is flammable. Store away from heat, sparks, and open flames. Contact hazard: wear gloves when handling.
3. Scott's Tap Water Substitute Blueing Solution (Fisher Scientific, Pittsburgh, PA, USA).
4. Ethanol gradient: 70% (v/v in purified $H_2O$), 95%, and 100% ethanol. Prepare fresh ethanol solutions weekly, or sooner if staining more than 20 slides/day, or if the ambient humidity is greater than 40%.
5. Ethyl alcohol, absolute, 200 proof for molecular biology. Ethanol is flammable. Store away from heat, sparks, and open flames. Do not ingest. Contact hazard: wear gloves when handling.

6. Purified water (Type I reagent grade water).
7. Xylene (Mallinckrodt Baker, Inc., Phillipsburg, NJ, USA). Xylene vapor is harmful or fatal, use with appropriate ventilation, discard in appropriate hazardous waste container. Xylene is flammable, store and use away from heat, sparks, and open flame. Contact hazard: wear gloves when handling.
8. Protease inhibitors (Complete Protease Inhibitor Cocktail Tablets, Roche, Indianapolis, IN, USA).

## 2.3. Laser Capture Microdissection

1. AutoPix™ or Veritas™ LCM system (Molecular Devices).
2. CapSure™ Macro LCM Caps (Molecular Devices) (*see* **Note 3**).
3. 500 μl Microcentrifuge Tubes: Safe-Lock Eppendorf™ tubes (Catalog No. 22 36 361-1, Brinkmann Instruments, Inc., Westbury, NY, USA); or MicroAmp™ 500 μl Thin-walled PCR Tubes (Catalog No. 9N801-0611, Applied Biosystems, Foster City, CA, USA).
4. Extraction buffer for cellular constituent of interest.

## 3. Methods

The protocols described below illustrate (1) frozen section sample preparation, (2) hematoxylin and eosin (H&E) tissue staining, and (3) automated LCM. Alternative tissue preparation methods, such as ethanol or formalin fixation with paraffin embedding, are acceptable for RNA and DNA analysis respectively *(30)*.

## 3.1. Frozen Tissue Sectioning

Surgical or biopsy material should be embedded directly in a cryopreservative solution and frozen immediately, or snap frozen in liquid nitrogen as soon as the specimen is procured. Prompt preservation of the sample limits protein and RNA degradation because of protease and RNase activity, respectively.

1. The samples should be cut at 2–15 μm thickness (5–8 μm is optimal) on plain, uncharged, pre-cleaned glass microscope slides (*see* **Note 4**).
2. Position the tissue section near the center of the slide, avoiding a 5-mm margin from either edge of the slide if using a Veritas™ instrument (*see* **Note 5**).
3. Do not allow the tissue section to dry on the slide at room temperature. Place the slide directly on dry ice or keep the slide in the cryostat at –20°C or colder until the slides can be either stained and microdissected, or stored at –80°C (*see* **Note 6**).

### 3.2. H&E Staining for Frozen Tissue Sections (for Downstream Proteomic or Genomic Analysis)

Classic tissue-staining protocols allow visualization of the tissue or cells of interest with a standard inverted light microscope. Most cellular staining protocols are compatible with LCM (*see* **Note 7**) and fluorescent stains are compatible with fluorescence equipped AutoPix™ and Veritas™ systems (*see* **Note 8**). Effective microdissection is achieved by minimal fixation of the tissue in 70% ethanol followed by staining of the cellular constituents, with a final dehydration in an ethanol gradient. Complete dehydration of the tissue is necessary for minimizing the upward adhesive forces between the tissue section and the slide (*see* **Note 9**). RNA degradation may be minimized by limiting the duration of the staining procedures. Protein degradation may be minimized by adding protease inhibitors to the staining reagents, in addition to limiting the microdissection session to 1 h *(31)* (*see* **Note 10**).

1. Remove the frozen section slide from freezer and place on dry ice or directly into the 70% ethanol fixative bath.
2. Dip, or gently shake, the slide in each of the following solutions, for the time indicated. Blot the slide on absorbent paper in-between each solution to prevent carryover from the previous solution.
3. 70% ethanol fixative            3–10 s
4. dH$_2$O                          10 s
5. Mayer's hematoxylin             15 s
6. dH$_2$O                          10 s
7. Scott's tap water substitute    10 s
8. 70% ethanol                     10 s
9. Eosin-Y (optional)              3–10 s
10. 95% ethanol                    10 s
11. 95% ethanol                    10 s
12. 100% ethanol                   30 s–1 min
13. 100% ethanol                   30 s–1 min
14. Xylene                         30 s–1 min
15. Xylene                         30 s–1 min
16. Allow the stained slide to air dry briefly. Proceed directly with microdissection. The stained slide should not be re-frozen (*see* **Note 11**).

### 3.2.1. H&E Staining for FFPEs (for Downstream Genomic Analysis only)

Formalin-fixed tissue has extensive protein crosslinking because of the formation of methylene bridges during formalin fixation. The crosslinks limit the amount of protein that can be retrieved from a tissue section. In general, microdissected FFPE tissue is best suited to genomic analyses.

Paraffin-embedded tissue sections must be deparaffinized before staining. Xylene acts as a solvent to remove the paraffin. Rehydration of the slide allows staining of the tissue elements.

| | | |
|---|---|---|
| 1. | Xylene | 5 min |
| 2. | Xylene | 5 min |
| 3. | 100% ethanol | 30 s |
| 4. | 95% ethanol | 30 s |
| 5. | 70% ethanol | 30 s |
| 6. | $dH_2O$ | 10 s |
| 7. | Mayer's hematoxylin | 15–30 s |
| 8. | $dH_2O$ | 10 s |
| 9. | Scott's tap water substitute | 10 s |
| 10. | 70% ethanol | 10 s |
| 11. | Eosin-Y (optional) | 3–10 s |
| 12. | 95% ethanol | 10 s |
| 13. | 95% ethanol | 10 s |
| 14. | 100% ethanol | 30 s–1 min |
| 15. | 100% ethanol | 30 s–1 min |
| 16. | Xylene | 30 s–1 min |
| 17. | Xylene | 30 s–1 min |

18. Allow the stained slide to air dry briefly. Proceed directly with microdissection (*see* **Note 11**).

### 3.3. Automated LCM (AutoPix™ System)

Cover slips and mounting media are not compatible with microdissection. In addition, a lack of immersion fluids on any of the optics prevents refraction of light from the tissue image. Thus, the color and detail of a given tissue stain is lost as the stained slide dries. Automated LCM methods overcome this fault by capitalizing on the index refraction of a wet tissue, allowing index-matched images to be saved and annotated for cell selection, permitting more accurate identification of cellular morphology during cell selection. Visualization of cellular morphological features, and the tissue in general, is achievable with image magnifications up to 40× through optical and color digital imaging. The resolution of the stitched images is constant, but the area of the images changes with magnification, permitting precise areas of tissue to be annotated for microdissection.

Annotation software coupled with the index-matched image permits single point dissection, line dissection, or polygon dissection (*see* **Fig. 3**). Algorithm-based, cell image recognition software on the AutoPix™ platform enhances the LCM technology. The algorithm is based on texture, morphology, size, color, and contrast of the tissue permitting automated cell selection in addition to automated microdissection *(28)*.

                                           *Rodriguez et al.*

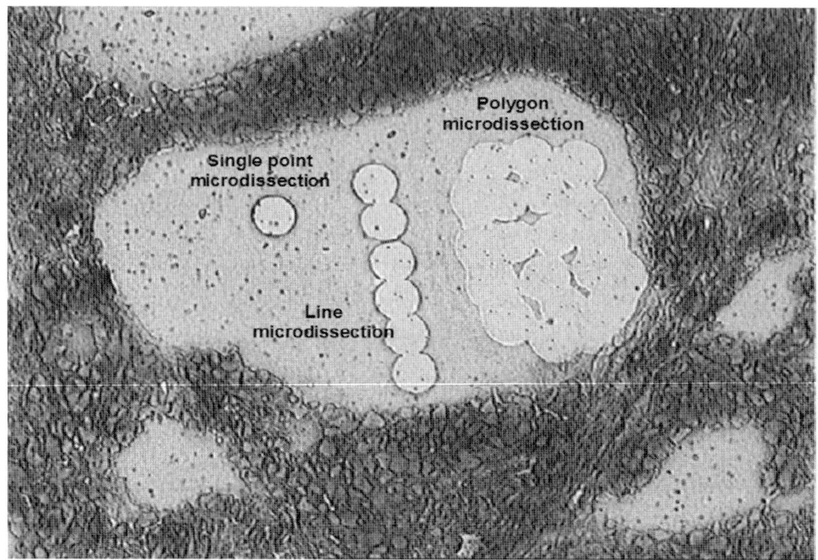


Fig. 3. Annotation styles for microdissection. Annotation software permits single point, line, or polygon area selection for microdissection. The single point style, with a resolution of 3–5 µm, permits microdissection of single cells. The line style is often used to microdissect single cell layers such as endothelial cells, whereas the polygon style is effective for microdissection of relatively large areas of homogeneous cells.

The AutoPix™ visualization system does not include oculars because of the enclosed system configuration. Instead, a PAL-format color camera permits visualization of the slide as a "roadmap image" for determining the target area of microdissection (*see* **Fig. 1**) *(28)*.

### 3.3.1. AutoPix™ Instrument Procedure for Microdissection

1. The AutoPix™ requires a warm-up period of approximately 1 h if the instrument is turned off.
2. Click on the AutoPix icon. Enter user name and password.
3. Load slide(s) and caps. Click OK on the Materials loading screen.
4. Click and drag the red camera box on the roadmap image to an area of interest.
5. Focus the live video image.
6. Right click on the roadmap image. Change image size by clicking on "Zoom 400%."
7. Acquire static images.

   a. Click on the red rectangle (region of interest) tool in the toolbar.
   b. Click and drag on the roadmap image to draw a region of interest.
   c. Click on the region of interest tool to close tool.

8. Click and drag the red camera box on the roadmap image to an area that will represent the center of the cap.

9. Click and drag the cap onto the slide.

10. Focus the laser. Adjust the target beam until the beam reaches the point of sharpest intensity and most concentrated light with little or no "halo-ing" or coronas. The laser should now be focused for any objective (*see* **Notes 12** and **13**).

11. Fire the laser by double clicking on the live video. Check polymer wetting. Observe the wetted polymer after the laser is fired. Firing the laser pulse causes the polymer to melt in the vicinity of the laser pulse. There should be a distinct clear circle surrounded by a dark ring (*see* **Fig. 4**) (*see* **Note 14**).

12. Suggested settings are power 70 mW and pulse 2500 µs. The size of the wetted polymer may be adjusted by changing the power and duration of the laser. These

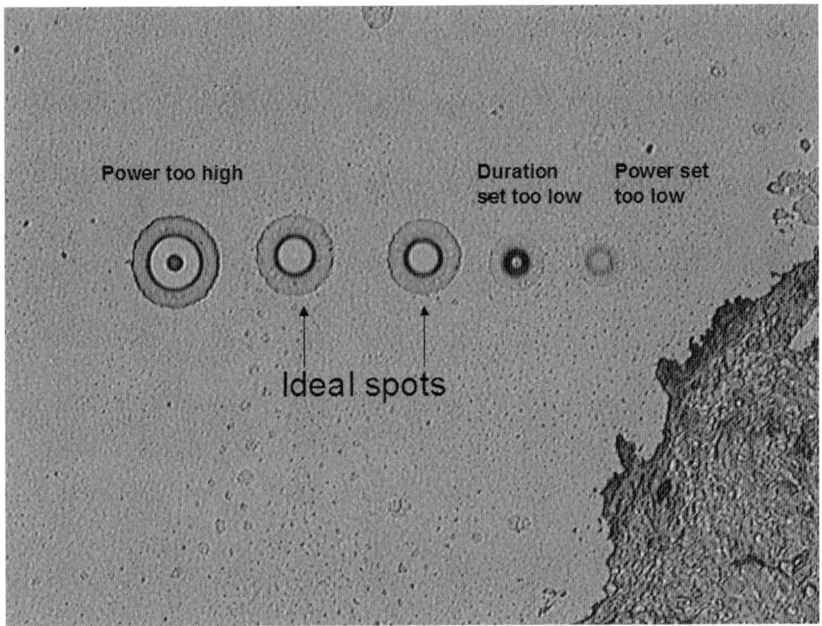


Fig. 4. Proper polymer wetting is essential for efficient microdissection. Adequate power, duration, proper tissue thickness, and cap placement on the tissue are parameters effecting the melting, or wetting, of the polymer. An ideal wetted polymer appears as a distinct dark ring with a clear center (arrows in figure). The presence of a dark area in the center of the spot indicates the power and/or duration settings are too high (leftmost spot). Inadequate power and duration result in failure of the laser to melt the polymer, creating a spot that appears as a gray, fuzzy circle (right most spot). Adequate power with inadequate duration may lead to spots with minute diameters as shown by the spot second from the right.

settings can be adjusted to customize the melted polymer spot to the type and thickness of the tissue to be dissected.

13. Measure the spot size diameter by clicking the mouse on one side of the wetted polymer spot and then clicking on the opposite side of the spot (do not click and drag). The measured diameter is displayed in the spot size window.

14. Select cells for microdissection by annotating the static images for the cells of interest. The annotation tools are (1) single point, (2) line, or (3) polygon.

15. Capture cells by right clicking on the static image and selecting "capture selected cells." The capture process will continue until all annotated areas for that particular image have been captured. Each annotated static image must be captured sequentially, as each static image represents a unique area for microdissection.

16. Click and drag the cap to the QC station to view the captured cells embedded in the polymer. The cap may be returned to the original slide for capturing additional cells, to another slide, or to the unload tray.

17. After miscrodissecting all cells of interest, click and drag the cap to the unload tray.

18. Click Instrument/Open instrument door.

19. Remove cap from the instrument. If viewing the cap reveals debris or non-specific tissue adhesion, the said material may be removed by gently blotting the polymer surface with an adhesive-type notepad (regular adhesive, not super sticky adhesive).

20. Insert the polymer end of the cap into the top of a 500-µl microcentrifuge tube. The sample is now ready for extraction of the desired components, or the cap-tube assembly may be stored for extraction at a later date (*see* **Notes 15 and 16**).

21. Repeat **steps 7–17** for additional static images.

22. To microdissect a new slide, click Start new session:

    a. Instrument/Start new session.
    b. Unload slides and **ALL** caps in the unload tray.
    c. Insert a new stained slide for microdissection and repeat **steps 2–17**.

23. To exit and shutdown:

    a. File/Exit
    b. Save images if needed.
    c. Unload all caps in the unload tray, remove unused caps, remove slides.
    d. Do not turn the instrument power off. The instrument power may remain on for daily operation eliminating the need for an initial warm-up period.

### 3.3.1.1. DUAL LASER ABLATION/CUTTING AND AUTOMATED LCM (VERITAS™ SYSTEM)

The Veritas™ combines robotics and optical scanning software with a UV laser and a near-infrared laser for automated microdissection of selected cells by one of three options: (1) traditional capture mode with a near-infrared laser, (2) Class IV UV cutting laser for rapid outlining and removal of large areas of tissue (*see* **Table 1**), or (3) UV laser for ablation, by photo volatilization, of unwanted or contaminating tissue (*see* **Note 17**). The software navigation tools are similar to the AutoPix™; therefore, only the UV cutting laser procedure will be described in detail for the Veritas™ instrument.

PEN membrane glass slides are required for UV laser cutting. The UV cutting laser option is designed to cut through the tissue and membrane in segments, with short sections, or tabs, that are not cut. These tabs prevent the cut tissue from curling away from the rest of the tissue section. A polymer cap is then placed on the cut area, and the near-infrared capture laser is used to tack the polymer to the cut tissue, enabling the complete removal of the cut tissue and underlying membrane from the slide.

1. The Veritas™ requires a warm-up period of approximately 1 h if the instrument is turned off.
2. Click on Veritas icon. Enter user name and password.
3. Load slide(s) and caps. If using membrane slides, click on membrane slide button.
4. Click OK on the Materials loading dialog box.
5. Double click on the roadmap image to position the red camera box over an area of interest.
6. Focus live video image.

**Table 1**
**Correlation Between UV Laser Cut Area and Number of Laser Capture Microdissection (LCM) Capture Pulses in the Veritas™ System *(7)*.**

| Spot size (μm) | Captured Area (μm$^2$) | Pulses | Captured Area/ Pulses (μm$^2$/pulse) |
|---|---|---|---|
| 15.1 | 328469.919 | 1509 | 217.674 |
| 15.1 | 462041.1 | 2102 | 219.81 |
| 15.1 | 656118.484 | 2988 | 219.584 |
| 30.1 | 216228.093 | 240 | 900.95 |
| 30.1 | 372304.058 | 409 | 910.279 |
| 30.1 | 634759.581 | 705 | 900.368 |

Average Captured Area/Pulse for a 15.1 μm spot size = 219.022 μm$^2$/pulse
Average Captured Area/Pulse for a 30.1 μm spot size = 903.866 μm$^2$/pulse

Using the cutting laser tools, circular regions were cut out of a tissue section. The "ruler tool" was used to measure each region's area. Using 15- and 30-μm laser spot-sizes, number of LCM capture pulses was correlated to UV laser cut area. The size and number of capture pulses is reflective of the number of cells to be obtained for analysis. These calculations allow the user to determine the area of a cut sample that is equivalent to a specified number of laser pulses or number of cells.

7. Click Tools/Cutting Laser to view the cutting laser dialog box.
8. Adjust the UV laser power by selecting either low, medium, or high power. Fine tune the laser with the slider bar in the cutting laser dialog box.
9. Click "Pulse" to fire the UV laser.
10. Set the spot size for the UV cutting laser. Click "spot" on the cutting laser toolbar and enter a defined value or perform a cut, use the ruler tool to measure the width of the cut, click "set spot" and enter the width measured with the ruler. Click "OK" to close the dialog box.
11. For membrane slides, set the cut properties for each capture group. Click on a capture group. Click "enable tabs." Enter values for the minimum number of tabs, spacing and size. Click "OK."
12. Click on the live video image.
13. Select the "cut and capture" tab on the microdissection dialog box. The knife icon allows free form outlines to be drawn for microdissection. The circle icon allows circular areas to be outlined for microdissection (*see* **Fig. 2**). The ablation tab allows areas to be marked for photo volatilization with the UV laser.
14. Annotate the live video image with the cutting tool of choice (*see* **Fig. 5**).

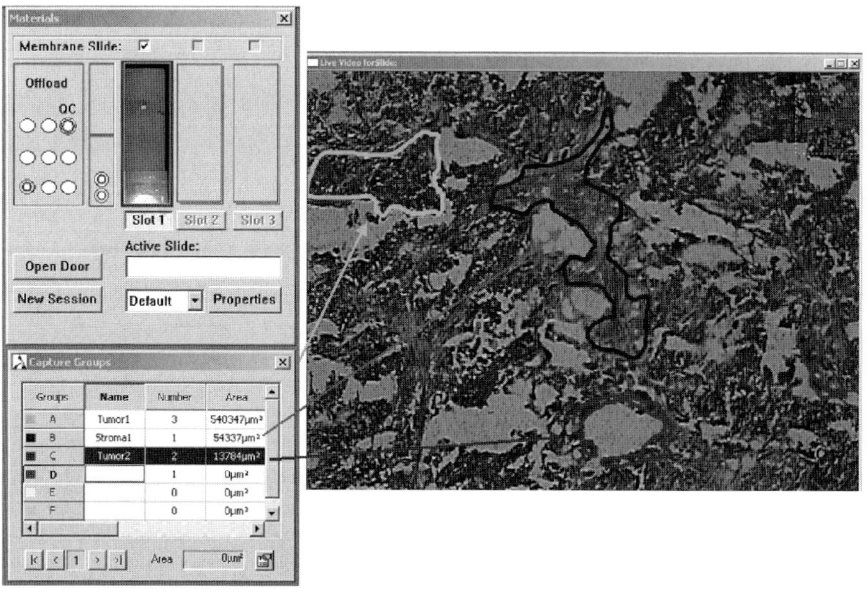

Fig. 5. Live video microdissection. Veritas™ instruments have the capacity to microdissect directly from the live video image. Various cell populations may be annotated during a single session, matched to a capture group and microdissected on separate Laser Capture Microdissection (LCM) caps. These features are beneficial for morphological differentiation of cells in heterogeneous sections as well as the ability to select multiple cell populations for microdissection from one tissue section.

15. Right click on the live video and select "cut selected groups."
16. A cap will be placed on the tissue after the UV laser cuts the desired area(s). Once the cap is in place, the near IR laser will tack the polymer to the cut tissue.
17. Click and drag the cap to the QC station or to the unload tray.
18. Click Instrument/Open door.
19. Remove the cap. Insert the polymer end of the cap into the top of a 500-µl microcentrifuge tube. The sample is now ready for extraction of the desired components or the cap-tube assembly may be stored for extraction at a later date (*see* **Note 15**).
20. Repeat **steps 11–18** for additional areas/caps.
21. To microdissect a new slide, click Start new session:

    a. Instrument/Start new session.
    b. Unload slides and **<u>ALL</u>** caps in the unload tray.
    c. Insert a new stained slide for microdissection and repeat **steps 2–18**.

22. To exit and shutdown:

    a. File/Exit
    b. Save images if needed
    c. Unload all caps in the unload tray, remove unused caps, remove slides
    d. Do not turn the instrument power off. If the power is turned off. The instrument power may remain on for daily operation eliminating the need for an initial warm-up period.

### 3.3.2. Storing Samples for Downstream Analysis

Microdissected cells for protein analysis may be stored at –80°C before extraction. Microdissected cells for DNA analysis may be stored desiccated at room temperature up to 1 week before extraction. Samples for RNA analysis should not be stored before extraction. Condensation in the microcentrifuge tube during storage may be a potential source of RNAse contamination. RNA extraction should be performed immediately after microdissection. Store extracted RNA samples at –80°C before amplification.

## 4. Notes

1. The near-infrared laser diode has a maximum laser output of 100 mW. The short laser pulse widths utilized, the low laser power levels required, the absorption of the laser pulse by the polymer and dye, and the long elapsed time between laser pulses combine to prevent the experimenter from depositing any significant amount of heat at the tissue surface, which may affect later laboratory analysis. The Class IV UV cutting laser, on the contrary, ablates the cells in the vicinity of the laser pulse. The UV cutting tool is designed for microdissection of larger

areas in which the cell number in the vicinity of the laser cut does not contribute significantly to the total cellular area captured.

2. Optimal tissue thickness for microdissection is 5–8 μm. Tissue sections less than 5 μm may not provide a full cell thickness, necessitating microdissection of more cells for a given assay. Tissue sections thicker than 8 μm may not microdissect completely, leaving integral cellular components adhering to the slide. Ethanol fixed tissue sections may also be used for proteomic analysis, but a larger number of microdissected cells may be required for adequate sensitivity in downstream assays as compared with frozen sections.

3. CapSure HS caps are often utilized in microdissection of tissue for RNA analysis. A 12-μm rail on the surface of the polymer prevents the polymer from touching the tissue except in the vicinity of the laser pulse. The HS caps are designed with an extraction device, allowing extraction buffer to contact the polymer within a centrally designated area. These features limit any potential RNA contamination from surrounding cells. CapSure HS caps may be used successfully for DNA or protein extraction. By contrast, CapSure Macro caps are placed in direct contact with the tissue and are not equipped with an extraction device. Any cellular material on the surface of the polymer of a Macro cap will be available for extraction.

4. Lung tissue, or other tissue with a thin, open architecture, may be cut on charged or silanized slides to prevent the tissue from non-specifically adhering to the polymer during microdissection. In general, coated slides are not used for microdissection because of the increased adhesive forces between the tissue and the slide. Effective microdissection is a balance between three adhesive forces: (1) maximizing downward adhesive forces between the polymer and the tissue, (2) minimizing lateral adhesive forces between the cells, and (3) minimizing upward adhesive forces between the slide and the tissue. PEN membrane frame slides may be used in automated LCM systems to microdissect living "in vitro" cells or to avoid contact between the polymer cap and tissue during microdissection. To microdissect living cells, (1) cells are grown in the PEN membrane frame slides™, (2) growth media is removed, (3) the slide is inverted and the cap is placed on the membrane, (4) the area of interest is outlined with near-infrared laser capture pulses, which melt polymer onto the membrane, and (5) the cap is removed, shearing selected cells away from the membrane frame slide.

5. The AutoPix™ stage has a capacity for 1–3 slides and the microdissectable area of the slide is approximately 19 × 45 mm. In contrast, the Veritas™ microdissectable area is constrained by a 5-mm margin from either edge of the slide, and a 7-mm margin from the bottom edge. This constraint is due to the metal edge of the frame slides and the need to design the instrument so the UV laser will not fire in the vicinity of the metal frame. This constraint affects glass slides as well as membrane slides in the Veritas.

6. Avoid repeated temperature fluctuations with the cut frozen sections. Store the frozen section slides at –80°C until the time of microdissection. Repeated

fluctuations in temperature may cause the tissue to adhere more tightly to the slide, limiting the effectiveness of microdissection.

7. Examples of LCM compatible stains are H&E, Methylene Blue, Wright–Giemsa or Toluidine Blue. Eosin staining of the cytoplasm is not necessary for visualization of cells during microdissection. Minimal staining times, in whichever staining protocol is utilized, limit potential protein alterations because of contact with the staining reagents. Selection of tissue-staining protocols should be based on compatibility with the downstream analysis to be performed with the microdissected tissue.

8. AutoPix™ and Veritas™ instruments equipped with fluorescent modules incorporate xenon lamps with blue, green, and red filter cubes. Additional filter cube positions are available for end-user modifications. Blue filter cubes use 455–495 nm excitation wavelengths with emission greater than 510 nm. Green filter cubes use 503–547 nm excitation wavelengths, with emission greater than 565 nm. The red filter cube excitation wavelengths are 590–650 nm with emission greater than 667 nm. Immunostained tissue is also compatible with LCM *(32–34)*.

9. Prolonging the 100% ethanol dehydration steps of the staining protocol to a maximum of 5 min may enhance tissue dehydration, maximizing microdissection efficiency. Skin tissue, cartilage, and samples prepared on charged slides may be difficult to microdissect. Additional slide or tissue treatments such as incorporation of glycerol slide coatings or modified staining protocols with glycerol may be required *(35)*. The following is one such approach for frozen tissue sections as adapted from *(35)*: Stain the slide in Mayer's hematoxylin for 30 s. Rinse in dH$_2$Ofor 15 s. Fix the slide in 70 and 95% ethanol solutions for 10 s each. Rinse in dH$_2$Ofor 10 s. Place the slide in Scott's Tap Water Substitute (Blueing) solution for 15 s. Dehydrate the slide in 70% ethanol for 2 min. Soak the slide in 3% glycerol in phosphate-buffered saline for 5–10 min. Dehydrate in two solutions of 100% ethanol, first solution for 10 s and in second solution for 1 min. Clear the slide in two changes of xylene or xylene substitute for 1 min each. Allow the slide to air dry before proceeding to microdissection.

10. Protease and/or phosphatase inhibitors may be added to compatible staining solutions *(17)*. Complete Protease Inhibitor Cocktail tablets are water soluble. For protease inhibitor addition to the 70% ethanol solution, dissolve the tablet in 15 ml of *dH$_2$O*, and then add 35 ml ethanol. Limiting the time from staining to completion of microdissection also ensures preservation of cellular constituents. Frozen section samples for RNA and protein analysis should be stained and microdissected within 1 h.

11. A drawback of the manual (PixCell™) system is the inability to microdissect directly from an index-matched image of the tissue. This is not an issue with the AutoPix™ or Veritas™ platform. Cells for microdissection are selected through annotation software directly from an index-matched, static stitched image, or in the case of the Veritas™, from the live video image or a static image. Index-matched images may be obtained with either system by rewetting the tissue with a drop of xylene before microdissection. It is imperative that the slide

be completely dry before cap placement for microdissection because xylene dissolves the polymer on the LCM cap.

12. The laser should not be refocused when changing objectives or spot sizes. It is only necessary to focus the laser once for each initial cap placement and any time the cap is repositioned on the tissue. Microdissection may be performed with any suitable laser spot size and a 4×, 10×, or 20× objective (PixCell) or a 2×, 10×, 20×, or 40× objective (Veritas).

13. Satellite laser spots are a phenomenon noted on the AutoPix™ platform. The satellite laser spot is a second surface reflection of the laser as the laser beam propagates through the optics. The laser is reflected from the back (second) surface of a coated optic. The coating is required to allow the laser to change direction in the optical path. Imperfections in the coating allow a portion of the laser beam to pass through the coating and reflect back to the viewer from the second optical surface. The power of the satellite spot is too low to melt the polymer and does not interfere with efficient microdissection. The satellite spot is a by-product of the light amplification system and was not designed as a system component, but the satellite spot is helpful for accurately focusing the primary laser spot. If the satellite spot is in focus, the primary laser spot will be in focus.

14. The dark ring produced by pulsing the laser is a combination of migration of the dye and changes in the thickness of the polymer wall at the site of the laser pulse (*see* **Fig. 4**). These changes in the polymer permit visualization of the melted polymer. The black ring should be sharp in appearance with a clear center. This pattern indicates (1) proper laser focusing, (2) adequate laser operation, and (3) acceptable performance of the CapSure polymer. A "fuzzy" ring could indicate improper focusing of the laser, uneven placement of the CapSure cap on the tissue, or inadequate power and/or duration of the laser pulse. The first step in troubleshooting a poorly wetted polymer spot is repositioning the cap on the tissue. Often times, the cap is crooked or uneven in relation to the tissue. The second step in resolving poorly wetted polymer spots is refocusing the laser. The third action to correct poor polymer wetting is adjustment of the power and duration. Increase the laser power and the duration and fire another laser test pulse. Observe the wetted polymer for the appropriate appearance. If the above steps fail to resolve the problem, discard the cap and repeat the process with a fresh cap.

15. Microdissected tissue for protein analysis may be stored directly on the LCM cap at –80 °C before protein extraction/cell solubilization. Microdissected tissue for RNA analysis should be immediately placed in RNA extraction buffer. Extracted RNA may be stored at –80°C before amplification. Microdissected samples for DNA analysis may either be extracted immediately or be stored at room temperature in a desiccated environment.

16. Assuming an average epithelial cell diameter of 7 µm and a 30-µm laser spot size, the operator can expect to collect, on average, 5–6 cells per laser pulse. Using this information, it is possible to estimate the number of cells captured based on

the number of laser pulses counted during microdissection. The number of laser pulses is automatically counted on the laser toolbar, fire option. For reverse phase protein microarrays, approximately 30,000–50,000 cells are required per 30 µl of extraction buffer. A more concentrated protein solution yields more robust signal : noise on a reverse phase protein microarray *(36)*. Mass spectrometric protein sequencing has been successfully performed with 20,000 microdissected cells.

a. 30 µm laser spot size: Number of pulses × 5 = total cells captured
b. 15 µm laser spot size: Number of pulses × 3 = total cells captured
c. 7.5 µm laser spot size: Number of pulses × 1 = total cells captured

17. The live video image may be used for cell annotation, rather than relying on stitched images, with the Veritas™ instrument. This feature is particularly useful in conjunction with the UV cutting laser, eliminating the need to acquire static images of the desired tissue area.

## Acknowledgments

The authors thank Meghan Liel, Edwin Posadas, and Chiara Pazzagli for thorough evaluations of the automated microdissection systems in our laboratory. The coral specimen was kindly provided by Shawn McLaughlin, NOAA/National Ocean Service/Cooperative Oxford Lab.

## References

1. Wulfkuhle, J.D., Aquino, J.A., Calvert, V.S., Fishman, D.A., Coukos, G., Liotta, L.A., and Petricoin, E.F., 3rd. (2003) Signal pathway profiling of ovarian cancer from human tissue specimens using reverse-phase protein microarrays. *Proteomics* **3**, 2085–90.
2. Sheehan, K.M., Calvert, V.S., Kay, E.W., Lu, Y., Fishman, D., Espina, V., Aquino, J., Speer, R., Araujo, R., Mills, G.B., Liotta, L.A., Petricoin, E.F., 3rd, and Wulfkuhle, J.D. (2005) Use of reverse phase protein microarrays and reference standard development for molecular network analysis of metastatic ovarian carcinoma. *Mol Cell Proteomics* **4**, 346–55.
3. Petricoin, E.F., Bichsel, V., Calvert, V., Espina, V., Winters, M., Young, L., Belluco, C., Steinberg, S., Trock, B., Lippman, M., Fishman, D., Sgroi, D., Munson, P., Esserman, L., Liotta, L.A. (2005) Mapping molecular networks using proteomics: A vision for patient-tailored combination therapy. *J Clin Oncol* **23**, 3614–21.
4. Liotta, L.A. and Kohn, E.C. (2001) The microenvironment of the tumour-host interface. *Nature* **411**, 375–9.
5. Liotta, L.A. and Kohn, E.C. (2002) Stromal therapy: the next step in ovarian cancer treatment. *J Natl Cancer Inst* **94**, 1113–4.

6. Michener, C.M., Ardekani, A.M., Petricoin, E.F., 3rd, Liotta, L.A., and Kohn, E.C. (2002) Genomics and proteomics: application of novel technology to early detection and prevention of cancer. *Cancer Detect Prev* **26**, 249–55.

7. Espina, V., Wulfkuhle, J.D., Calvert, V.S., VanMeter, A., Zhou, W., Coukos, G., Geho, D.H., Petricion, E.F., 3rd, and Liotta, L.A. (2006) Laser-capture microdissection. *Nat Protoc* **1**, 586–603.

8. Cowherd, S.M., Espina, V.A., Petricoin, E.F., 3rd, and Liotta, L.A. (2004) Proteomic analysis of human breast cancer tissue with laser-capture microdissection and reverse-phase protein microarrays. *Clin Breast Cancer* **5**, 385–92.

9. Liotta, L.A. and Clair, T. (2000) Cancer. Checkpoint for invasion. *Nature* **405**, 287–8.

10. Liotta, L. and Petricoin, E. (2000) Molecular profiling of human cancer. *Nat Rev Genet* **1**, 48–56.

11. Hunter, T. (2000) Signaling–2000 and beyond. *Cell* **100**, 113–27.

12. Hanahan, D. and Weinberg, R.A. (2000) The hallmarks of cancer. *Cell* **100**, 57–70.

13. Jeong, H., Tombor, B., Albert, R., Oltvai, Z.N., and Barabasi, A.L. (2000) The large-scale organization of metabolic networks. *Nature* **407**, 651–4.

14. Oltvai, Z.N. and Barabasi, A.L. (2002) Systems biology. Life's complexity pyramid. *Science* **298**, 763–4.

15. Ma, X.J., Wang, Z., Ryan, P.D., Isakoff, S.J., Barmettler, A., Fuller, A., Muir, B., Mohapatra, G., Salunga, R., Tuggle, J.T., Tran, Y., Tran, D., Tassin, A., Amon, P., Wang, W., Enright, E., Stecker, K., Estepa-Sabal, E., Smith, B., Younger, J., Balis, U., Michaelson, J., Bhan, A., Habin, K., Baer, T.M., Brugge, J., Haber, D.A., Erlander, M.G., and Sgroi, D.C. (2004) A two-gene expression ratio predicts clinical outcome in breast cancer patients treated with tamoxifen. *Cancer Cell* **5**, 607–16.

16. Zang, L., Palmer Toy, D., Hancock, W.S., Sgroi, D.C., and Karger, B.L. (2004) Proteomic analysis of ductal carcinoma of the breast using laser capture microdissection, LC-MS, and 16O/18O isotopic labeling. *J Proteome Res* **3**, 604–12.

17. Lechpammer, M. and Sgroi, D.C. (2004) Laser Capture Microdissection: a rising tool in genetic profiling of cancer. *Expert Rev Mol Diagn* **4**, 429–30.

18. Emmert-Buck, M.R., Bonner, R.F., Smith, P.D., Chuaqui, R.F., Zhuang, Z., Goldstein, S.R., Weiss, R.A., and Liotta, L.A. (1996) Laser capture microdissection. *Science* **274**, 998–1001.

19. Bonner, R.F., Emmert-Buck, M., Cole, K., Pohida, T., Chuaqui, R., Goldstein, S., and Liotta, L.A. (1997) Laser capture microdissection: Molecular analysis of tissue. *Science* **278**, 1481–3.

20. Berman, D.M., Wang, Y., Liu, Z., Dong, Q., Burke, L.A., Liotta, L.A., Fisher, R., and Wu, X. (2004) A functional polymorphism in RGS6 modulates the risk of bladder cancer. *Cancer Res* **64**, 6820–6.

21. Bichsel, V.E., Liotta, L.A., and Petricoin, E.F., 3rd, (2001) Cancer proteomics: from biomarker discovery to signal pathway profiling. *Cancer J* **7**, 69–78.

22. Chuaqui, R., Cole, K., Cuello, M., Silva, M., Quintana, M.E., and Emmert-Buck, M.R. (1999) Analysis of mRNA quality in freshly prepared and archival Papanicolaou samples. *Acta Cytol* **43**, 831–6.

23. Chuaqui, R., Vargas, M.P., Castiglioni, T., Elsner, B., Zhuang, Z., Emmert-Buck, M., and Merino, M.J. (1996) Detection of heterozygosity loss in microdissected fine needle aspiration specimens of breast carcinoma. *Acta Cytol* **40**, 642–8.

24. Miura, K., Bowman, E.D., Simon, R., Peng, A.C., Robles, A.I., Jones, R.T., Katagiri, T., He, P., Mizukami, H., Charboneau, L., Kikuchi, T., Liotta, L.A., Nakamura, Y., and Harris, C.C. (2002) Laser capture microdissection and microarray expression analysis of lung adenocarcinoma reveals tobacco smoking- and prognosis-related molecular profiles. *Cancer Res* **62**, 3244–50.

25. Xu, H., Chaturvedi, R., Cheng, Y., Bussiere, F.I., Asim, M., Yao, M.D., Potosky, D., Meltzer, S.J., Rhee, J.G., Kim, S.S., Moss, S.F., Hacker, A., Wang, Y., Casero, R.A., Jr., and Wilson, K.T. (2004) Spermine oxidation induced by Helicobacter pylori results in apoptosis and DNA damage: implications for gastric carcinogenesis. *Cancer Res* **64**, 8521–5.

26. Xu, B.J., Caprioli, R.M., Sanders, M.E., and Jensen, R.A. (2002) Direct analysis of laser capture microdissected cells by MALDI mass spectrometry. *J Am Soc Mass Spectrom* **13**, 1292–7.

27. DiFrancesco, L.M., Murthy, S.K., Luider, J., and Demetrick, D.J. (2000) Laser capture microdissection-guided fluorescence in situ hybridization and flow cytometric cell cycle analysis of purified nuclei from paraffin sections. *Mod Pathol* **13**, 705–11.

28. (2002) *User Guide AutoPix Automated Laser Capture Microdissection System.* Arcturus Engineering: Mountain View, CA. p. 5

29. (2004) *User Guide Veritas Microdissection Instrument.* Arcturus Bioscience, Inc: Mountain View, CA.

30. Gillespie, J.W., Best, C.J., Bichsel, V.E., Cole, K.A., Greenhut, S.F., Hewitt, S.M., Ahram, M., Gathright, Y.B., Merino, M.J., Strausberg, R.L., Epstein, J.I., Hamilton, S.R., Gannot, G., Baibakova, G.V., Calvert, V.S., Flaig, M.J., Chuaqui, R.F., Herring, J.C., Pfeifer, J., Petricoin, E.F., Linehan, W.M., Duray, P.H., Bova, G.S., and Emmert-Buck, M.R. (2002) Evaluation of non-formalin tissue fixation for molecular profiling studies. *Am J Pathol* **160**, 449–57.

31. Simone, N.L., Remaley, A.T., Charboneau, L., Petricoin, E.F., 3rd, Glickman, J.W., Emmert-Buck, M.R., Fleisher, T.A., and Liotta, L.A. (2000) Sensitive immunoassay of tissue cell proteins procured by laser capture microdissection. *Am J Pathol* **156**, 445–52.

32. Fend, F., Emmert-Buck, M.R., Chuaqui, R., Cole, K., Lee, J., Liotta, L.A., and Raffeld, M. (1999) Immuno-LCM: laser capture microdissection of immunostained frozen sections for mRNA analysis. *Am J Pathol* **154**, 61–6.

33. Kinnecom, K. and Pachter, J.S. (2005) Selective capture of endothelial and perivascular cells from brain microvessels using laser capture microdissection. *Brain Res Brain Res Protoc* **16**, 1–9.

34. Buckanovich, R.J., Sasaroli, D., O'Brien-Jenkins, A., Botbyl, J., Conejo-Garcia, J.R., Benencia, F., Liotta, L.A., Gimotty, P.A., and Coukos, G. (2006) Use of immuno-LCM to identify the in situ expression profile of cellular constituents of the tumor microenvironment. *Cancer Biol Ther* **5**, 635–42.

35. Agar, N.S., Halliday, G.M., Barnetson, R.S., and Jones, A.M. (2003) A novel technique for the examination of skin biopsies by laser capture microdissection. *J Cutan Pathol* **30**, 265–70.
36. Espina, V., Mehta, A.I., Winters, M.E., Calvert, V., Wulfkuhle, J., Petricoin, E.F., 3rd, and Liotta, L.A. (2003) Protein microarrays: Molecular profiling technologies for clinical specimens. *Proteomics* **3**, 2091–100.

# 6

# Tissue Microarrays

*An Overview*

## Rajiv Dhir

## Summary

Traditionally, screening for new markers involves using a slide from each of several different patients. A more efficient way is to have one slide that contains several minute specimens, one from each patient. These slides are prepared by transferring paraffin tissue cores from many "donor" blocks to one "recipient" block. Each slide cut from this recipient block is called a tissue microarray (TMA) slide. It can have various histological types of disease that need to be compared or can have the same histological type but different behavior (e.g., responders versus non-responders, etc.). TMAs are ideal for efficient screening of prospective biomarkers by a variety of different mechanisms including immunohistochemistry, fluorescence in situ hybridization of nucleic acids (FISH) and RNA in situ hybridization. Selection and number of cases from patient subsets in a given microarray slide is amenable to statistical modeling to enhance analysis of results. In addition, different microarrays can be constructed to answer different scientific questions. The microarrays can also be produced from retrospective paraffin blocks of well-characterized cases, with clinical follow-up. The TMA slides can be "whole-slide" imaged. This provides a mechanism to share results of experiments with other investigators. There are also ongoing efforts to generate software tools for automated analysis of TMA localization data. There has also been a significant body of work done to standardize data capture, thus facilitating subsequent exchange of information. The preferred current mechanism is to use an "XLM"-based data capture and transfer. There have also been efforts to create "frozen" TMAs. This has been attempted using "donor" frozen tissues embedded in OCT compound. These samples are then arrayed into a recipient OCT block. The presence of OCT can sporadically interfere with certain assays. However, it does provide a novel mechanism for high-throughput evaluation of frozen tissue, with corresponding visualization of tissue morphology.

From: *Methods in Molecular Biology, vol. 441: Tissue Proteomics: Pathways, Biomarkers, and Drug Discovery*
Edited by: B.C.-S. Liu © Humana Press, Totowa, NJ

**Key Words:** Paraffin Tissue Microarray; TMA; XLM data exchange; frozen array; array design.

## 1. Introduction

The conventional mechanisms for localization and expression studies of paraffin tissue-based analysis of large numbers of tumors are slow and tedious. Traditionally, screening for a new marker involves using a slide from each of several different patients. A more efficient way is to have one slide that contains several minute specimens, one from each patient *(1)*. Each slide thus created is called a paraffin tissue microarray (TMA) (*see* **Fig. 1**). The TMA can have various histological types of tumors that need to be compared or can have the same histological type but different behavior (e.g., metastatic, non-metastatic, radiosensitive, radio resistant, high grade, low-grade, etc.). A single slide cut from a microarray block can have representative samples from different tumor types *(2)*. These slides are prepared by transferring paraffin block tumor cylindrical cores from many "donor" blocks to one "recipient" block. These

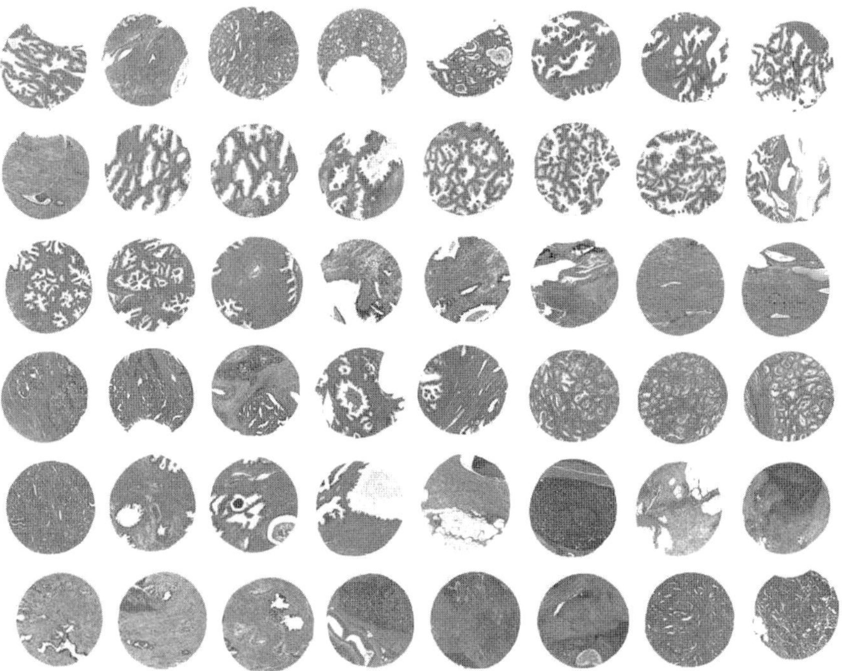

Fig. 1. A low power view of a multi-tumor paraffin tissue microarray (TMA).

cores are neatly arranged in identifiable positions in rows and columns (hence the name "microarrays"). The location of each tissue in the donor block is cataloged. Appropriate quality control mechanisms need to be in place, with quality checks every few levels.

TMAs are ideal for efficient screening of prospective biomarkers by a variety of different mechanisms including immunohistochemistry, fluorescence in situ hybridization of nucleic acids (FISH) and RNA in situ hybridization *(3,4)*. These techniques help in tumor profiling, rapid screening of gene amplifications in cancer, verifying in vivo differential expression of genes identified by cDNA arrays, and identifying prognostic and diagnostic markers. This approach saves time for researchers and decreases the waste of slides and reagents in the screening phase of biomarker evaluation.

The power of the TMA technique is the capability of performing a series of analyses of thousands of specimens in a parallel fashion. Sections cut from the array allow parallel detection of DNA, FISH, mRNA, or IHC targets in each of the hundreds of specimens in the array. This allows consecutive analyses of a large number of molecular markers and construction of a database of correlated genotype or phenotypic characteristics of the tumor type being evaluated.

Selection of a number of cases from patient subsets in a given microarray slide is amenable to statistical modeling to enhance analysis of results. Different microarrays can be constructed to answer different scientific questions. The microarrays can also be produced from archival material, using paraffin blocks of already characterized tumors with clinical follow-up. This would provide a rapid evaluation of clinically well-characterized tumors. The viability of this approach has been tested with prostate carcinomas, renal cell carcinomas, and other tumors *(5,6)*.

Most TMA blocks will be utilized for a multitude of research projects. The initial design of the TMA might be done with a specific project in mind. However, it is important to keep the design flexible and capable of addressing a range of research questions. There should be significant planning regarding the design of the TMA. Important things to consider are maximizing utilization of available resources and incorporating appropriate numbers of cases to address potential statistical issues. It is not possible to design the perfect TMA, which would be capable of handling all statistical as well as research questions; however, the TMA capable of addressing a multitude of research issues can be designed with input from the research community and biostatisticians (*see* **Note 1**).

There are a variety of TMAs that have been constructed by different research groups. These TMAs help address different questions. Because many of these have extensive data annotation, it is possible to pool available information and provide larger cohorts of cases for research studies.

### 1.1. Progression-Based TMA

This is the most frequent type of TMA constructed. This type will have cores that represent the entire spectrum of disease, starting with tissues from normal individuals and ending with tissues obtained from distant metastases of patients who are refractory to treatment. Ideally, a progression TMA will include cases without cancer (true normal controls), pre-malignant disorders, and histologically normal tissues from areas adjacent to a cancer, histologically normal tissues from areas distant to a known cancer, cancers at various stages and grades, as well as regional and distant metastases. A single subject may contribute with several of these tissue cores *(1)*. Understanding the distinction between normal tissue obtained from an individual without known cancer and normal appearing tissue obtained from an individual with known cancer is crucial in the interpretation of results because there are always concerns about the phenomenon of "field effect" in many tissue types *(7)*.

### 1.2. Tumor TMA

This is similar to the progression-based TMA. However, this TMA consists only of needle cores from foci of tumor. This TMA is useful for assessing the frequency of marker expression throughout the spectrum of tumor differentiation. The major pitfall that needs to be recognized is subjectivity related to tumor grading *(8)*. It is critical to utilize well-defined, universally accepted criteria. It might be useful to design quality assurance (QA) exercises at the outset, to ensure uniformity of grading, as well as involve more than one pathologist to grade the tumors. There should be a process of resolution of discordant grading, with a discussion to reach consensus on specimens that receive discordant grading assignments.

### 1.3. Outcomes-Based TMA

This TMA is designed to provide a case set with extensive clinical information and follow-up. Outcome-based TMAs address questions pertaining to candidate cancer prognostic markers. Cases with the longest clinical follow-up and best-documented outcome measures are the most useful in this type of array. The main concern in building this type of TMA lies in assuring that the number of cases available for arraying has enough statistical power to address outcome-related issues. Another major concern with outcome TMAs is the purity of the annotating data. Poor data quality can significantly compromise the value of these TMAs. Statistical considerations are another important aspect of construction of outcome TMAs. There must be a sufficient number of cases available and included, in order to provide any meaningful assessment using outcome arrays.

### 1.4. Tumor Heterogeneity TMA

This type of TMA is designed to evaluate heterogeneity of expression of a marker in the same lesion (*see* **Note 2**). This TMA therefore requires many cores from the same subject in order to provide a meaningful assessment of heterogeneity of expression. This type of TMA may include samples from the centre and the periphery of the tumor, as well as samples from metastatic foci.

### 1.5. Consecutive Cases TMA

This is the most frequent type of TMA encountered in the literature. This TMA encompasses most/all available cases in a particular repository. The primary use of this type of TMA is to evaluate the frequency of expression of a given antigen in a particular disease/cancer type. This approach may be the best design modality for smaller repositories because it provides an opportunity to evaluate all available cases. This approach, however, has a major limitation: many projects address questions that do not require the significant numbers of cases included in a consecutive cases TMA. The most frequent reason for exclusion is that these cases do not address the experimental question.

### 1.6. Specialty TMA

A speciality TMA is designed with a specific project in mind. These TMAs suffer from the major drawback that they require a significant investment of time and effort for a single project. An alternative solution to the customized TMA construction approach is a "modular" approach, where existing TMAs can be used to complement each other. Another important concern is the utilization of scarce resources. This is specifically significant when evaluating cases with extensive follow-up information as well as foci of metastatic disease. Metastatic tissue samples are not readily available in department of pathology archival materials, and they should be used only for validated projects.

### 1.7. TMA from Biopsy Material

Most TMAs are constructed using paraffin blocks derived from large resection specimens. Needle biopsy specimens represent another interesting and important tissue material available for research purposes. Biopsy specimens are especially interesting for diagnostic markers. An elegant strategy for creating TMAs from needle biopsies is described in a recent publication *(9)*. TMA blocks designed with this strategy can generate up to 200 paraffin sections.

### 1.8. Tissue Culture Cells Array

A TMA can be constructed from cultured cell lines. These cell line samples can serve as control samples because they have been extensively used in the literature and their profile for the marker of interest might be well documented *(10–12)*. The cell line pellets are processed in a manner similar to that for routine surgical pathology tissue. This is important because it eliminates any variability related to fixation protocols with reference to immunohistochemical reactivity and sample viability. Use of a resin-based approach has also been described *(13)*. This technique provides excellent morphology and minimizes variability of marker expression related to fixation protocols.

### 1.9. Preservation Issues

There is evidence that staining intensities in stored TMA sections deteriorate over a period of time *(14)*. Because the TMA has limited representation of each case, this problem could significantly impact marker evaluation and provide false-negative results (*see* **Note 2**). Either cutting a fresh section at the time of evaluation or using other mechanisms to preserve antigenicity can resolve this problem. Cutting a fresh section every time can significantly impact the number of sections available from one TMA block, because trimming the block results in the wasting of valuable tissue. Alternative solutions consist of storing TMA sections in a vacuum, in an environment consisting of nitrogen gas, as well as at –80 °C.

There is no common consensus on this issue. Future studies and advances in technology might lead to a universally acceptable solution.

### 1.10. Data Annotation Issues

The design, construction, and use of paraffin TMAs present a variety of challenges in the realm of informatics *(15)*. Tissue banks intending to provide TMAs must store relevant patient, accession, and specimen information and pathologic descriptions down to the block level, as well as be able to associate that data with a given core. Tissue array production systems should provide user interfaces that allow for consistent description of tissue regions. In addition, it is ideal to have an image of a slide taken from the block with annotation of the area(s) from where the core(s) were taken.

Currently, pathologists evaluate immunohistochemical or in situ hybridization studies. However, many groups are actively evaluating automated methods. The data extracted from the evaluation can be conveniently divided into two main areas: quality control data and experimental data. Quality control data include information on the integrity of the core (e.g., presence/absence of tumor in the current section), as well as any internal controls that may relevant;

this will depend strongly on the type of tissue and the nature of the stain. Experimental data include the presence and extent of staining, as well as the distribution/pattern of staining (i.e., cell membrane versus nuclear) in some cases. Consistency of results reporting overall cores and across multiple arrays is of extreme importance. The user interface should facilitate this process by including evaluation guidelines or standard images.

Slides made from microarrays can be examined directly through the microscope or indirectly through digital imaging systems. Digital imaging systems for microscopy have improved greatly over the past several years, and several systems provide virtual microscopes that allow the user to examine the entire array on a single, low power screen and to zoom onto a single cell in a single core. Digital imaging allows storage of the raw data (image) and the ability to collaborate over a network. Digital imaging can also display two or more physically separate cores side by side for evaluation or superimpose the same core (with different staining) to visualize the spatial relationship of different protein or gene expression. The computational analysis of digital images will be an increasingly important technique in the coming years.

TMA experiments are data intensive, requiring substantial information to interpret, replicate, or validate. Recently, an open-access Tissue Microarray Data Exchange Specification has been released that allows TMA data to be organized in a self-describing XML document annotated with well-defined common data elements *(16)*. Although this specification provides sufficient information for the reproduction of the experiment by outside research groups, its initial description did not contain instructions or examples of actual implementations, and no implementation studies have been published. Data elements can be extracted from an Excel database using a transformative Perl script. TMA databases conforming to the Tissue Microarray Data Exchange Specification can be merged with other TMA files, expanded through the addition of data elements, or linked to data contained in external biological databases.

It is important to note data aggregation and storage issues because large-scale collaborative efforts require similar nature of databases.

### *1.11. Frozen TMAs*

There have also been efforts to create "frozen" TMAs. This has been attempted using "donor" frozen tissues embedded in an OCT compound. These samples are then arrayed into a recipient OCT block *(17)*. An adaptation of this process has been utilized for brain specimens *(18)*.

Some publications have reported the successful use of OCT arrays for DNA, RNA, and protein analyses *(17)*. This approach may have significant advantages over the use of paraffin tissue-based TMA technology for the assessment of some genes and proteins by improving both qualitative and quantitative results. However, the presence of OCT can interfere with certain assays. Nonetheless, the frozen tissue TMA does provide a novel mechanism for high-throughput evaluation of frozen tissue, with corresponding visualization of tissue morphology.

## 2. Materials

The construction of a TMA is a multi-step process. The multiple steps are listed below and discussed in detail in the *Methods* section.

1. Identify tissue type and case type.
2. Design matrix.
3. Donor paraffin blocks.
4. Fresh haematoxylin and eosin (H&E) slides from blocks listed in *(3)*.
5. Marking of slides obtained in *(4)*.
6. Tissue Microarrayer.
7. Recipient block.
8. Slides from TMA block.
9. QA of the TMA slides.

## 3. Methods

1. *Identification of tissue type and case type from which the TMA will be constructed:* The initial step is to decide the tissue type and disease type from which the TMA will be constructed. This is primarily a decision driven by the research initiative requiring the creation of a TMA. There needs to be statistical input regarding the size of the case set to be included in the TMA. This first step allows the investigators to address the feasibility of the project. In addition, this also has subsequent implications in the design of the TMA.

2. *Design matrix:* There are a variety of ways of creating a TMA matrix.

   a. Marker cores should be inserted into the TMA from tissue types not present in the array (*see* **Note 2**).
   b. Many investigators use contiguous rows and columns in the architectural design of the TMA. We prefer to use a matrix with 4 × 4 squares and intervening spacer rows. Spacing of 0.8 mm is recommended between core samples to create the most efficient array. This is needed to aid in proper orientation of the TMA as well as assist in automated imaging. This design approach is illustrated in **Fig. 2**. This design approach helps address issues related to "drift and shift" seen during the cutting of a TMA block. The drift and shift refers to small and large changes respectively in orientation of the tissue cores because of technical issues related to sectioning of the TMA block.

3. *Identification of the blocks to be punched to create the TMA:* The cases needed for the TMA being constructed are first selected. The next step is to identify the appropriate blocks from the selected cases.

| MM-001 | MM-002 | MM-003 | MM-004 | | MM-006 | MM-011 | MM-045 | MM-046 |
|---|---|---|---|---|---|---|---|---|
| MM-005 | MM-006 | MM-007 | MM-008 | | MM-047 | MM-048 | MM-049 | MM-050 |
| MM-009 | MM-010 | MM-011 | MM-012 | | MM-051 | MM-052 | MM-053 | MM-054 |
| MM-013 | MM-014 | MM-015 | MM-016 | | MM-001 | MM-002 | MM-003 | MM-004 |
| | | | | | | | | |
| MM-018 | MM-019 | MM-021 | MM-022 | | MM-065 | MM-066 | MM-067 | MM-068 |
| MM-055 | MM-056 | MM-057 | MM-058 | | MM-069 | MM-070 | MM-071 | MM-072 |
| MM-059 | MM-060 | MM-060 | MM-061 | | MM-032 | MM-073 | MM-074 | MM-075 |
| MM-062 | MM-062 | MM-063 | MM-064 | | MM-076 | MM-077 | MM-030 | MM-038 |

Fig. 2. Design template used at the University of Pittsburgh.

   a. It is advised that each block selected be evaluated for the thickness of tissue in the block (*see* **Note 3**). This can be done visually as well as using radiological techniques. We have generated pilot data using the Faxitron machine for assessing the thickness of tissue in the paraffin block.

4. *Marking of slides to localize the site to be punched:* The blocks should have fresh H&E sections cut. The sections are then provided to the pathologist to mark the lesional areas to be punched for creating the TMA.

   a. *Tissue Microarrayer to create the TMA:* There are many vendors of equipment for creating paraffin TMAs. The most popular method for TMA construction is the manual method. A variety of machines are available to automate some or all components of TMA construction (*see* **Note 4**).

5. *Preparation of the TMA block:*

   a. It is recommended that at least 2–6 replicate tissue samples from the same block be inserted in the array, as some may be lost during processing. The consensus is to have 3–4 replicate samples to eliminate any impact of heterogeneity of expression.

   b. The core size generally used ranges from 0.6 to 2.0 mm. The larger core sizes are preferred for tissue types with space between the cells/lot of stroma. The

2.0-mm core size is generally used for TMA and blocks being constructed from lung tissue and stromal/adipose tissue.

c. There might be instances where a block is depleted and might not have adequate depth of lesional tissue. We have used a process of stacking multiple donor needle cores from the same tissue type in the same recipient hole. This is a practical way of utilizing tissue from blocks that have been depleted.

d. Post construction treatment or Block Curing is done after construction of the TMA block. The TMA block is warmed in a 37 °C incubator for 30 min or in a 45 °C incubator for 20 min. Light and even pressure is applied to the surface of the block with a clean glass slide to level core heights. Blocks are immediately transferred to an ice bath or the freezer to cool before sectioning.

e. The blocks are cooled and then placed in the microtome and faced before the construction of the TMA. This sets the microtome for the proper alignment of the block for the sectioning phase of the array.

6. *Cutting of the TMA block to make slides from the TMA block:* Cured TMA blocks are sectioned by one of two methods (*see* **Note 5**).

a. The conventional method involves floating a paraffin ribbon on a heated water bath.

b. The alternative method uses the Paraffin Sectioning Aid System. The block is first faced if using the Paraffin aid system. An adhesive film window is applied to the surface of the block using a roller system. The roller ensures coverage of all the tissue samples and no air bubbles.

c. Slides cut the conventional way are floated on a water bath and picked up on Superfrost Plus slides. Slides are dried in a conventional oven (55–60 °C) for several hours or overnight.

7. *QA of the TMA block:* TMA slides are serially sectioned in quantities of 50–100 slides at one time. The TMA block is generally cut to generate many slides at a time to avoid tissue wastage during facing of the block. Every 25th section is stained, using H&E, to confirm histologic characteristics of the tissue samples.

## 4. Notes

The paraffin TMA provides a tool for high-throughput evaluation of localization initiatives. A large number of samples can be evaluated on a very limited number of slides. The data generated from the localization experiments can be linked to other databases with clinical information. This process provides a mechanism for rapid evaluation of a large number of cases. The data generated are also amenable to statistical evaluation and manipulation.

1. Appropriate statistical considerations are also critical when using the TMA approach for marker validation. It is important to have an appropriate number of cases available for evaluation to provide meaningful data. The case identification

should also be done keeping in mind and clinical variables that might need to be assessed as part of the project.

2. The technical problems associated with TMA use relate primarily to two major areas. First, there may be degradation of antigen as a result of oxidative influences during storage. Secondly, tumor heterogeneity may confound data aggregation using this high-throughput method. The conventional slide modality of localization efforts provides the opportunity to evaluate large amounts of tissue specimens. Any technical issues related to the staining process could significantly reduce the value of the data generated. It is very important to have appropriate QA/QC slides evaluated with the TMA sections. In addition, the TMA should have inbuilt positive controls (cell lines/other tissue types).

3. It is also important to keep in mind that the process of creation of the TMA can result in significant compromise of the lesional tissue in the selected block. It is therefore important to make sure that there are backup blocks for clinical diagnostic purposes in case additional studies might need to be performed in future.

4. The automated systems have a broad price range, depending on the degree of automation. In addition, some laboratories have had local initiatives to develop automated TMA construction *(19)*. As is expected, the equipment needed for the manual method is the least expensive **Fig. 3** shows a manual TMA machine. It is important for the laboratory embarking on an investment in this area to evaluate the machines. Each machine is different with different capabilities designed to

Fig. 3. A manual Tissue Microarray construction machine.

ease the process of TMA construction. The volume of work anticipated is an important criteria in this decision-making process. If the workload is projected to be relatively light, large investments in an automated system might not be justified. However, these are local decisions dependent on funding issues.

5. There is no consensus on the best process for sectioning the TMA blocks. Some investigators prefer the water bath protocol whereas others prefer the paraffin-sectioning aid system. Some investigators claim that the paraffin-sectioning aid system can impact immunohistochemical assessment.

## Acknowledgment

The author thanks The Cooperative Prostate Cancer Tissue Resource.

## References


1. Kononen, J., L. Bubendorf, A. Kallioniemi, M. Barlund, P. Schraml, S. Leighton, J. Torhorst, M. J. Mihatsch, G. Sauter, and O. P. Kallioniemi. 1998. Tissue microarrays for high-throughput molecular profiling of tumor specimens. *Nat Med* **4**: 844–7.
2. Eguiluz, C., E. Viguera, L. Millan, and J. Perez. 2006. Multitissue array review: A chronological description of tissue array techniques, applications and procedures. *Pathol Res Pract* **202**: 561–8.
3. Braunschweig, T., J. Y. Chung, and S. M. Hewitt. 2004. Perspectives in tissue microarrays. *Comb Chem High Throughput Screen* **7**: 575–85.
4. Moch, H., T. Kononen, O. P. Kallioniemi, and G. Sauter. 2001. Tissue microarrays: What will they bring to molecular and anatomic pathology? *Adv Anat Pathol* **8**: 14–20.
5. Dhir, R., J. Gilbertson, and M. J. Becich. 2001. Developments in tissue banking for the post-genome era. *Advances in Anatomic Pathology* **8**: 307–309.
6. Schraml, P., J. Kononen, L. Bubendorf, H. Moch, H. Bissig, A. Nocito, M. J. Mihatsch, O. P. Kallioniemi, and G. Sauter. 1999. Tissue microarrays for gene amplification surveys in many different tumor types. *Clin Cancer Res* **5**: 1966–75.
7. Chandran, U. R., R. Dhir, C. Ma, G. Michalopoulos, M. Becich, and J. Gilbertson. 2005. Differences in gene expression in prostate cancer, normal appearing prostate tissue adjacent to cancer and prostate tissue from cancer free organ donors. *BMC Cancer* **5**: 45.
8. Oyama, T., W. C. Allsbrook, Jr., K. Kurokawa, H. Matsuda, A. Segawa, T. Sano, K. Suzuki, and J. I. Epstein. 2005. A comparison of interobserver reproducibility of Gleason grading of prostatic carcinoma in Japan and the United States. *Arch Pathol Lab Med* **129**: 1004–10.
9. Datta, M. W., A. Kahler, V. Macias, T. Brodzeller, and A. Kajdacsy-Balla. 2005. A simple inexpensive method for the production of tissue microarrays from needle biopsy specimens: examples with prostate cancer. *Appl Immunohistochem Mol Morphol* **13**: 96–103.

10. Li, R., J. Ni, P. A. Bourne, S. Yeh, J. Yao, P. A. di Sant'Agnese, and J. Huang. 2005. Cell culture block array for immunocytochemical study of protein expression in cultured cells. *Appl Immunohistochem Mol Morphol* **13**: 85–90.

11. Montgomery, K., S. Zhao, M. van de Rijn, and Y. Natkunam. 2005. A novel method for making "tissue" microarrays from small numbers of suspension cells. *Appl Immunohistochem Mol Morphol* **13**: 80–4.

12. Moskaluk, C. A. and M. H. Stoler. 2002. Agarose mold embedding of cultured cells for tissue microarrays. *Diagn Mol Pathol* **11**: 234–8.

13. Howat, W. J., A. Warford, J. N. Mitchell, K. F. Clarke, J. S. Conquer, and J. McCafferty. 2005. Resin tissue microarrays: a universal format for immunohistochemistry. *J Histochem Cytochem* **53**: 1189–97.

14. Fergenbaum, J. H., M. Garcia-Closas, S. M. Hewitt, J. Lissowska, L. C. Sakoda, and M. E. Sherman. 2004. Loss of antigenicity in stored sections of breast cancer tissue microarrays. *Cancer Epidemiol Biomarkers Prev* **13**: 667–72.

15. Dhir, R., J. Gilbertson, and M. J. Becich. 2001. Tissue microarrays and data analysis: An informatician's dream. *Advances in Anatomic Pathology* **8**(6): 361–362.

16. Berman, J. J., M. Datta, A. Kajdacsy-Balla, J. Melamed, J. Orenstein, K. Dobbin, A. Patel, R. Dhir, and M. J. Becich. 2004. The tissue microarray data exchange specification: Implementation by the Cooperative Prostate Cancer Tissue Resource. *BMC Bioinformatics* **5**: 19.

17. Schoenberg Fejzo, M. and D. J. Slamon. 2001. Frozen tumor tissue microarray technology for analysis of tumor RNA, DNA, and proteins. *Am J Pathol* **159**: 1645–50.

18. Kylaniemi, M., M. Koskinen, P. Karhunen, I. Rantala, J. Peltola, and H. Haapasalo. 2004. A novel frozen brain tissue array technique: immunohistochemical detection of neuronal paraneoplastic autoantibodies. *Neuropathol Appl Neurobiol* **30**: 39–45.

19. Matysiak, B. E., T. Brodzeller, S. Buck, A. French, C. Counts, B. Boorsma, M. W. Datta, and A. A. Kajdacsy-Balla. 2003. Simple, inexpensive method for automating tissue microarray production provides enhanced microarray reproducibility. *Appl Immunohistochem Mol Morphol* **11**: 269–73.

# 7

# Frozen Protein Arrays

## Stephen M. Hewitt and Robert A. Star


### Summary

This chapter describes the rationale behind and means of construction of inexpensive, low to moderate throughput protein arrays. The method of construction is based on injection of analytes into a block of frozen optimum cutting temperature (OCT), the gel media used for frozen sections, and sectioned on a cryostat. The "array section" is applied to a nitrocellulose pad. Once on nitrocellulose, the array can be utilized in any fashion desired. The analytes can be any biologic sample including peptides, proteins, antibodies, cells, nucleic acids, or any other material that can tolerate freezing. This platform provides investigators a flexible inexpensive easy-to-fashion platform to create multiplex assays both in the number of samples analyzed and in the types of assays.

**Key Words:** Protein; antibody; array; cells; cell lines; frozen tissue; lysate; dilution curve.


## 1. Introduction

High-throughput detection of protein abundance can be performed in two modes: (1) An antibody array, whereby antibodies are spotted onto a two-dimensional support—either the bottom of an enzyme-linked immunosorbent assay (ELISA) well or slide, or by conjugating them to the three-dimensional surface of a bead, that are then incubated with a complex protein mixture. (2) A protein array (often called reverse phase arrays) where 10s to 1000s are arrayed to the surface of a glass slide, and an antibody is hybridized to the surface. Although bead-based approaches can be simplified for construction within the laboratory, the two-dimensional platforms require array printers that are expensive, not readily available, challenging to use, and require


From: *Methods in Molecular Biology, vol. 441: Tissue Proteomics: Pathways, Biomarkers, and Drug Discovery*
Edited by: B.C.-S. Liu © Humana Press, Totowa, NJ

specialized specimen handling. The significant *overhead* of these platforms requires a substantial commitment of resources (both physical and economic) for a fixed number of arrays that prevents most laboratories from adopting these high-throughput approaches, especially when the number of uses is limited.

To address these challenges, we devised a moderate-throughput array platform that is flexible, inexpensive, easy to construct, requires only a small amount of critical reagents, and can be easily configured to meet the needs and experimental design of any investigator. Specimens are embedded into a preformed frozen block of optimum cutting temperature (OCT) embedding material with minimal specimen volume and preparation. The specimens can include proteins from any source [in vitro translation, cell lysates, tissue lysates from sources including laser capture microdissection (LCM) samples] *(1)*. Alternatively, antibodies can be embedded, producing an antibody array rather than a protein array. If the investigator desires, other material can be embedded in the platform, including, but not limited to, nucleic acids and intact cells. Because the investigator has complete control of the analyte in the array, dilution curves or mixed platforms of different sources can be constructed. Once the array block is constructed, it is sectioned on a conventional cryostat and the array section can be applied to any substrate desired, most commonly nitrocellulose-based platforms for the remainder of the assay. Because multiple sections can be cut from a single array block, up to about 600 assays can be performed on the array (*see* **Note 1**).

## 2. Materials

### 2.1. Frozen Array Block

1. Mold and punch (frame in which to freeze OCT)
2. Pin array (array of pins that form the wells in the OCT)
3. Desktop freezer or cryostat (–20 to +4 °C temperature range)
4. Cryogel OCT (Instrumedics, Hackensack, NJ, USA) (*see* **Note 2**)
5. Freezer, –80 °C

### 2.2. Mold and Punch for the Array Block

The mold consists of a 12-mm thick aluminum mold containing a 12 × 19-mm rectangular hole. The punch is a 12 × 9 × 24-mm rectangular piece of aluminum (*see* **Note 2**).

### 2.3. Pin Array Mold

An array of pins is used to form wells in the frozen OCT (*see* **Fig. 1**). We constructed a 5 × 5 pin array of 23-G hollow pins spaced 2 mm apart that

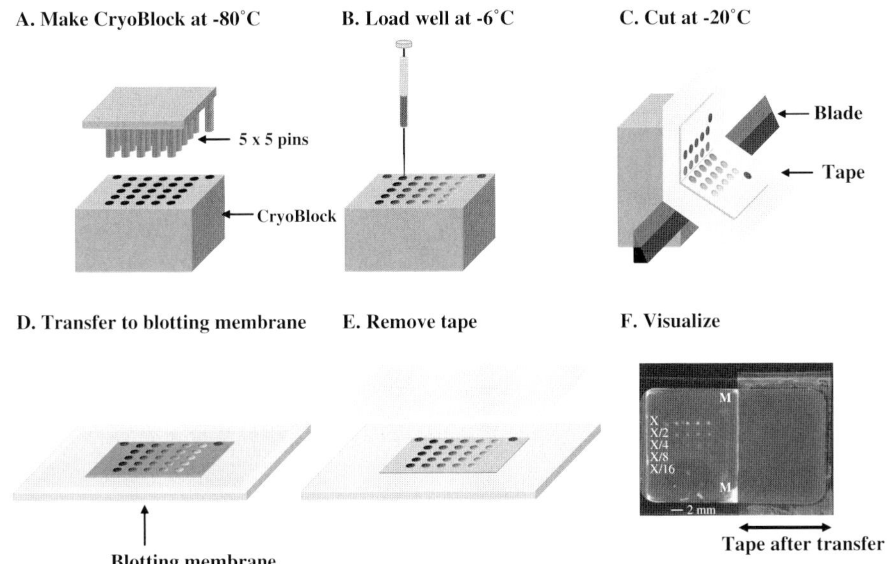



Fig. 1. Design, construction, and operation of frozen protein array. Used with permission of *Proteomics (1)*.

protruded 1 cm from an aluminum block (*see* **Fig. 1A**). Two additional pins were included to orient the frozen protein array (*see* **Note 3**). The individual pins should not be less than 500 µm in diameter; however, larger diameters can be utilized (requiring a larger volume of analyte per well). The pin's length should be uniform to produce wells of equivalent depth.

### 2.4. Analyte to be Arrayed

Any biospecimen that is not irreparably damaged by freezing can be arrayed (*see* **Note 4**). Ideally, the analyte should be as concentrated as possible. Depending on the ultimate design of the array, normalizing the concentration of the analytes may be desired. Some quantification of the analyte before arraying (protein concentration for example) is desirable, but not essential.

### 2.5. Filling the Frozen Array

1. Hamilton 10-µl glass syringe pipette
2. 50% Sucrose in water
3. 50% Colored Gelatin (Jell-O brand gelatin, or similar product, *see* **Note 5**) in room temperature water. Colored gelatin is preferred rather than substitution with biologic grade gelatins or agarose.

## *2.6. Sectioning the Frozen Array*

1. Cryostat
2. Scotch Magic Transparent Tape (3 M Corp)
3. Appropriate solid support (nitrocellulose FAST Slides, Schleicher and Schuell, Whatman Inc., Florham Park, New Jersey)

# 3. Methods

## *3.1. Construction of Array Mold*

1. The first step is casting an OCT mold in which the lysates will be deposited. Place the aluminum mold on a metal surface and chill to $-6\,°C$ using a benchtop freezer, cryostat, or with the aid of dry ice.
2. Fill the mold with OCT, avoiding bubbles (*see* **Note 6**). Do *not* allow the OCT to freeze completely.
3. Warm the OCT to $-6\,°C$ to soften the OCT.
4. Quickly inset the needle template to create the wells. Allow the block to freeze for 5 min. When frozen, remove the needle template by pulling straight up.

## *3.2. Organizing and Preparing Analytes for the Frozen Array*

1. Samples to be arrayed should be placed at $4\,°C$ (on ice). The total volume of the well is 3–5 µl (*see* **Note 7**).
2. In a chilled microfuge tube, combine analyte with a sucrose solution and gelatin to have a final concentration of 20% sucrose and 15% gelatin (*see* **Note 7**).
3. Mix the final solution well. Dilution samples should be diluted with water or appropriate sample buffer, and then add the sucrose and gelatin solution.

## *3.3. Construction of the Frozen Protein Array*

1. Warm the array block to $-6\,°C$ either in a cryostat or on a bench top cooling chamber (*see* **Note 8**).
2. Using a Hamilton glass syringe pipettor, inject the analyte into the appropriate well based on the array design. Be careful to avoid bubbles, but filling from the bottom of the well and backing out the needle. Continue and fill all the wells in the array design (*see* **Note 9**).
3. When all the wells have been filled, reduce the temperature of the freezing chamber to $-20\,°C$ or place the block on dry ice or in a $-80\,°C$ freezer (*see* **Note 10**).

## *3.4. Sectioning of the Frozen Protein Array*

1. Maintain the cyroarray block at $-20\,°C$ or below.
2. Mount the cryoarray block on an appropriate specimen holder for the cryostat. Typically, this is accomplished by using a small amount of OCT to glue the completed block to the holder.

3. Allow the block to freeze completely for 3–5 min. Once mounted, *face* the cryoblock until the surface of the array is completely flat and reasonable sections can be cut.

4. Set the microtome to the desired section thickness. Sections of 5–10 μm thickness are routine (*see* **Note 11**). Sections can be cut and directly placed on glass slides or nitrocellulose-based surfaces as if they were frozen tissue or TMA sections (*see* **Note 12**).

## *3.5. Use of Cryoarrays*

Once applied to the nitrocellulose surface, the resultant array can be assayed in an identical manner to the conditions routinely used for detection as a western blot or immunoblot (*see* **Note 13**).

## 4. Notes

1. The ultimate design of the array is up to the investigator. Some experimental designs will place a different specimen in each well, resulting in a multiplex dot-blot format. Alternatively, analytes can be diluted to create a dilution curve for more exact quantifications. A standard curve of biomaterial of known concentration can be included, and if combined with dilution curves of specimen, but utilized for exact quantification of analytes. The platform is compatible with functional assays including kinase assays, enzymatic processes and can be used to study protein–protein, protein–nucleic acid, and nucleic acid–nucleic acid interactions.

2. The mold and punch can be constructed by a biomedical instrumentation shop, but are simple enough to be constructed in the laboratory. In theory, larger arrays can be constructed, such as for double wide (brain) slides or raw nitrocellulose membranes, measuring up to 40 by 45 mm. However, the final frozen array should not be larger than what can be cut on a cryostat or the size of the final support surface. Currently, the array block mold and punch are not available commercially.

   An array map or template should be constructed on paper to aid in placing the correct analyte in the correct well. A "map as you go" approach will lead to frustration and failure. Depending on the goals of the array, it may be beneficial to quantify total protein per analyte before constructing the array, or this concentration can be determined from the array if a standard curve of bovine serum albumin (BSA) is included.

3. The pin array can be fabricated out of metal or constructed with needles in a plastic/modeling clay holder. Hollow needles are preferred so that a vacuum does not form when the pins are removed from the frozen OCT. Syringe needles of 23-G work well. The layout of the needles in the mold can be user defined, but should be asymmetric to include at least one locating pin. Needles can be

any reasonable spacing (~2 mm) to accommodate the total number of wells desired in the platform.

4. Frozen specimens should be maintained in a frozen state as long as possible before arraying, and when defrosted, only warmed sufficient to produce a liquid state for mixing. If LCM-derived specimens are used, they should be maintained either at 4 °C or frozen till ready for arraying. Even lysates derived from formalin-fixed paraffin embedded tissue should not be left at room temperature after extraction.

   Nucleic acids, either DNA or RNA, are compatible with the format. Any routine buffer for resuspension of the nucleic acids can be utilized. The same general approach for proteins can be applied for nucleic acids. Although we do not have experience with glycoproteins, carbohydrates, or other biologic materials in this platform, we are unaware of any limitations to their applications. Whole cells prepared as a suspension of cells in PBS or media (with or without serum or 10% DMSO) can be arrayed. The number of cells should be determined before arraying. Depending on the number of cells per milliliter, the density of cells per section can vary from near tissue-like appearance to a single cell per final section. Concentrations of $10^6$cells/ml are adequate to ensure cell profiles on every section. If desired, the cryosections can be applied to a glass slide rather than a nitrocellulose surface, resulting in a frozen cell microarray (CMA) akin to a frozen tissue microarray (TMA). Frozen TMAs can be constructed; however, they require the use of needles to remove the cores of frozen tissue from the donor block *(2)*.

5. Jell-O or other food product gelatins are utilized to impart color as well as provide additional sucrose and gelatin to mark the filled wells and to improve the bonding of the analyte to the side of the well.

6. OTC is the generic name of the freezing mediums for embedding tissue. Experiments with Tissuetech OCT (Sakura Finetek, Torrance, CA, USA) were suboptimal. Cryogel OCT is composed of water (80%), sucrose (10%), and proprietary material (10%). If necessary, OCT can be degassed with a vacuum. Do *not* allow the OCT to freeze completely.

7. It is essential that the final concentration of analyte and support matrix be 20% sucrose and 15% gelatin. Because the maximal cool temperature concentration of sucrose in solution is approximately 50% for sucrose and for gelatin, it should be planned that each sample will be diluted 35% in the arraying procedure.

8. The block is warmed to –6 °C for the introduction of the analytes. At this temperature, the cryoblock remains solid, and the analyte/sucrose/gelatin solution remains liquid for sufficient time to ensure injection of the solution into the well. Although the solution may freeze after it is injected into the well, this is dependent on the concentration of the analyte. At –6 °C, the block is not sufficiently rigid for cutting and must be chilled further.

9. Fill empty wells with a solution of either Cryogel OCT or 20% sucrose and 15% gelatin, as empty wells can impair sectioning.

10. Once the array is completed, it must be chilled to –20 °C for sectioning. It is essential that the specimens thoroughly freeze to a solid state. It is suggested that if the block is to be stored before or after sectioning, then it should be wrapped in foil and stored at –80 °C or colder. Do not store the cryoarray in the cryostat overnight or for any extended periods. Most cryostats have automatic defrost cycles. Although these cycles may not warm the cryostat chamber sufficiently to deform the cryoarray (temperatures between –4 and +4 °C), they are sufficient to result in damage to the analytes.

11. Although sections can be cut at any thickness, the thickness that will determine the final concentration of analyte on the array is based on the equation:

    Analyte mass $= \pi r^2 \cdot d \cdot [\text{analyte}]$

    where $r$ is the diameter of the well, $d$ is the thickness of the section, and [analyte] is the concentration. Hence, a 20-µm section will deposit twice as much analyte per spot on the final array as a 10-µm section. This flexibility can be of utility as it is a means of increasing the density of the analyte per spot.

12. Sections can be cut and directly placed on glass slides or nitrocellulose-based surfaces as if they were frozen tissue or TMA sections. However, there may be significant distortion of the array and some array spots may be lost. It is recommended that a piece of scotch tape be applied to the frozen block with light pressure (a simple brush used in routine sectioning of frozen sections is adequate). With the tape in place, a single section is cut. The tape with adherent frozen protein array section is applied to a nitrocellulose surface and allowed to dry at 4 °C for 3–4 h. The affinity of nitrocellulose is sufficient to bind the biomolecules, and the tape can simply be lifted from the slide. Alternatively, the Cryo-Jane Tape transfer system (Instrumedics) can be utilized to apply the sections to glass slides.

13. For any array platform to give accurate results, it is essential to have general knowledge of interactions between and platform and the assays being performed. Appropriate positive and negative control spots should be included. If the array platform is being used for immunodetection of a protein by antibodies of haptens, it is essential that a western blot be performed to demonstrate the specificity of the antibody being utilized, such that the antibody detects only a single band, or known and previously described bands. The antibody should be validated on samples similar to that arrayed in the frozen protein array. We note that a single band on western blotting provides a common level of validation on the quality of the Ab, but it does not necessarily predict how it will function on other platforms.

    The array platform does not replace gel-based blotting methods (western, northern, Southern, and such), but rather with a validated assay of specificity allows a high-throughput and quantitative approach to these gel-based approaches. The goal is to apply more samples or dilutions to the format than can be accomplished with a standard gel format. There is the added benefit of being able to directly control the concentration of analyte assayed beyond the capacity of gel-based approaches, which are subject to overloading, within individual wells due to inaccurate quantitation.

This method is covered by US Patent No. 6,951,761 *(3)* and may not be used for commercial products without appropriate license from the DHHS/NIH. Use of this method for research applications is permitted and encouraged.

## Acknowledgments

The authors thank Takehiko Miyaji and Peter Yuen for their technical assistance in developing this method and useful comments on this manuscript. The authors thank Lance Liotta for his advice and insights into applications of this platform.

## References

1. Miyaji T, Hewitt SM, Liotta LA, and Star RA. Frozen protein arrays: A new method for arraying and detecting recombinant and native tissue proteins. *Proteomics* 2002, **2**: 1489–93.
2. Schoenberg Fejzo, M and Slamon DJ. Frozen tumor tissue microarray technology for analysis of tumor RNA, DNA, and proteins. *Am. J. Pathol.*, 2001, **159**(5): 1645–50.
3. Star RA, Miyaji T, Hewitt SM, and Liotta LA. Measurements of multiple molecules using a cryoarray, inventors. US Patent 6,951,761.

# 8

# Reverse Phase Protein Microarrays for Theranostics and Patient-Tailored Therapy

Virginia Espina, Julia Wulfkuhle, Valerie S. Calvert, Lance A. Liotta, and Emanuel F. Petricoin III

## Summary

Although the genome provides information about the somatic genetic changes existing in the tissue and underpins pathology, it is the proteins that do the work of the cell and are functionally responsible for almost all disease processes. Moreover, many diseases such as cancer are a manifestation of deranged cellular protein molecular networks and cell-signaling pathways. These pathways contain a large and growing collection of drug targets, governing cellular survival, proliferation, invasion, and cell death. Thus, the promise of proteomics resides in the study of molecules that extend beyond correlation to causality. The clinical utility of reverse phase protein microarrays, a new technology invented in our laboratory, lies in its ability to generate a functional map of known cell-signaling networks or pathways for an individual patient obtained directly from a biopsy specimen. This patient-specific circuit diagram provides key information that identifies critical nodes or pathways that may serve as drug targets for individualized or combinatorial therapy through the quantification of phosphorylation states of proteins. Using this technique, the entire cellular proteome is immobilized on a substratum with subsequent immunodetection of the phosphorylated, or activated, state of cell-signaling proteins. The results of which pathways are "in use" can then be correlated with biological and clinical information and serve as both a diagnostic and a therapeutic guide: thus providing a "theranostic" endpoint.

**Key Words:** Cancer; laser capture microdissection; microarray; molecular profiling; protein; proteomics; tissue heterogeneity; theranostics.

From: *Methods in Molecular Biology, vol. 441: Tissue Proteomics: Pathways, Biomarkers, and Drug Discovery*
Edited by: B.C.-S. Liu © Humana Press, Totowa, NJ

## 1. Introduction

Protein microarrays, in the simplest definition, are immobilized protein spots on a substratum that allows for the effective multiplexed analysis of at least several analytes *(1–6)*. The individual protein spots may be heterogeneous or homogeneous in nature. They may consist of whole cell lysates, or phage lysates, an antibody, a nucleic acid, body fluid, drug, or recombinant protein *(2,3,5–13)*. These immobilized bait molecules are detected by probing the protein microarray with a signal-generating molecule such as a tagged antibody, a labeled ligand, or even a tagged serum or cell lysate. The tagging molecule generates a pattern of positive and negative spots. The image can be generated from any number of colorimetric, flourimetric, radiometric, or evanescent means. The signal intensity of each spot is proportional to the quantity of applied tagged molecules bound to the bait molecule. The image is captured, analyzed, and correlated with biological information.

The focus of this method topic is on reverse phase protein microarrays (RPA), a method inherently different from the popular form of forward-phase arrays such as antibody microarrays. Unlike a forward phase approach, the RPA microarray format comprises an immobilized cellular lysate that is probed with a primary antibody. Signal amplification is independent of the immobilized protein, permitting the coupling of detection strategies with highly sensitive tyramide amplification chemistries *(14–17)*.

Protein microarrays have broad applications for discovery and quantitative analysis, with applications in drug discovery, biomarker identification, and molecular profiling of cellular material. The overall utility of the RPA lies in the ability to provide a "circuit" map of the state of multiple in vivo kinase-driven signal pathways from a tiny piece of tissue, and to provide crucial information about protein post-translational modifications, such as the phosphorylation states of these proteins *(18–20)*. These phosphorylation events ultimately reflect the "active" state of signal pathways and networks, and importantly, this information cannot be measured by gene microarrays. In essence, the commonly used "gene network analysis" is a misnomer because genes do not form function networks per se—it is the proteins that form the networks. Protein microarrays provide a view of the disrupted cellular machinery governing disease. Identification of critical nodes, or interactions, within these networks is a critical starting point for theranostics, or the use of diagnostic biomarkers that are also the therapeutic targets themselves: the beginning steps toward individualized therapy *(18,21–23)*. Characterization of protein cell-signaling pathways in tissue serves two critical purposes: (1) stratifying patients for therapy and (2) discovery of new therapeutic regimens for improving treatment outcomes.

The methods described herein encompass (1) preparation of a whole cell lysate from either cell culture or tissue samples, (2) protein lysate microarray printing, (3) immunostaining, and (4) microarray spot analysis.

## 2. Materials

### 2.1. Cellular Lysates

1. Cell culture, human or animal cellular sample. Satisfactory tissue samples for protein analysis are (1) frozen tissue sections, (2) ethanol-fixed, paraffin-embedded tissue, or laser capture microdissected (LCM) cell populations from the aforementioned tissues (*see* **Note 1**).
2. Microdissected samples on LCM caps with Safe-Lock Eppendorf Tubes, 0.5 ml (Brinkmann, Westbury, NY, USA, Cat. No. 22 36 361-1) or MicroAmp™ 500 ml Thin-walled Reaction Tubes (Perkin-Elmer Applied Biosystems, Foster City, CA, USA, Cat. No. N801-0611).
3. Cell lysis buffer: T-PER™ (Tissue Protein Extraction Reagent, Pierce, Rockford, IL, USA), 2× Tris-glycine sodium dodecyl sulfate (SDS)-loading buffer (Invitrogen, Carlsbad, CA, USA), 2-mercaptoethanol, 5 M NaCl (300 mM final conc.).
4. Protease inhibitors for *cell culture* samples: 200 mM PEFABLOC (AEBSF; 4-(2-aminoethyl)-benzenesulfonyl fluoride hydrochloride) (Roche Applied Science, Indianapolis, IN, USA, Cat. No. 1 585 916), 1 mg/ml aprotinin (5 µg/ml final conc., Roche, No. 1583794), 5 mg/ml pepstatin A (5 µg/ml final conc., Roche, No. 1524488), 5 mg/ml leupeptin (5 µg/ml final conc., Roche, No. 1529048).
5. Protease inhibitors for *tissue* samples: Protease Inhibitor Cocktail (Cat. No. P2714 Sigma Aldrich, St. Louis, MO, USA). Use 10 µl of protease inhibitor cocktail/ml of lysis buffer.
6. Phosphatase inhibitor: 1 mM orthovanadate (Sigma Aldrich)—please make this fresh every time.

### 2.2. Printing RPAs

1. Nitrocellulose-coated glass slides (FAST slides, Whatman Schleicher & Schuell Biosciences, Keene, New Hampshire, UK).
2. Microarray printing device (Aushon Biosystems 2470 Protein Arrayer, Burlington, MA, USA).
3. Cellular lysate sample (minimum volume 30 µl).
4. 96-well or 384-well polypropylene microtiter plate.
5. Purified water (type I reagent grade water).
6. 70% ethanol.
7. Desiccant (Drierite anhydrous calcium sulfate, W.A. Hammond Drierite Co. Xenia, OH, USA).
8. Ziplock-style plastic bags.

## 2.3. Immunostaining Microarrays

### 2.3.1. Microarray Pretreatment and Blocking

1. Reblot™ Mild Antigen Stripping solution 10× (Chemicon, Temecula, CA, USA) (*see* **Note 2**).
2. Phosphate-buffered saline (PBS) without calcium or magnesium.
3. I-Block™ Protein-Blocking Solution (Applied Biosystems).
4. Tween 20 (DakoCytomation, Carpinteria, CA, USA).

### 2.3.2. DakoCytomation Autostainer Immunostaining

Immunostaining may be performed manually if an Autostainer is not available.

1. Validated primary antibody of choice (*see* **Note 3**).
2. Biotinylated, species-specific secondary antibody.
3. DakoCytomation Autostainer (DakoCytomation).
4. Catalyzed Signal Amplification (CSA) kit (DakoCytomation).
5. Biotin-blocking system (DakoCytomation).
6. Antibody diluent with background reducing components (DakoCytomation).
7. Tris-buffered saline with Tween (DakoCytomation).
8. Liquid DAB+ (diaminobenzidene) (DakoCytomation).

### 2.3.3. Total Protein Blot Stain for Microarrays

1. Sypro Ruby Protein Blot stain (Cat. No. S-11791 Invitrogen/Molecular Probes).
2. Fixative solution: 7% acetic acid, 10% methanol in $dH_2O$ (acetic acid, glacial Cat. No. V193, Mallinckrodt; Methanol, absolute Cat. No. 17-5 Sigma Diagnostics).
3. Standard UV transilluminator or a laser scanner (FluorChem 8800, Alpha Innotech Imager, San Leandro, CA, USA) (*see* **Note 4**).
4. Sypro Red/Texas Red filter (490-nm longpass filter) or a 600-nm bandpass filter.

## 2.4. Microarray Spot Analysis

1. Spot analysis software of choice: examples are ImageQuant® (Amersham Biosciences), MicroVigene™ (Vigene Tech, Billerica, MA, USA) or P-SCAN [Peak quantitation with Statistical Comparative Analysis (http://abs.cit.nih.gov)].
2. High-resolution flatbed scanner (Epson® Perfection Scanner series 1640, Long Beach, CA or UMAX PowerLook 1120, Dallas, TX, USA).
3. Adobe® Photoshop software.

## 3. Methods

The protocols below describe: (1) protein lysate preparation, (2) protein (tissue) lysate microarray printing, (3) immunostaining, and (4) spot analysis. These protocols are uniquely designed for quantitative analysis of protein phosphorylation events in cellular lysates, with the concomitant analysis of the corresponding total (phosphorylated and non-phosphorylated) protein.

### 3.1. Protein Lysate Preparation

1. Samples may be cell culture lysates, microdissected tissue lysates, or other cellular lysates. These protocols are also compatible with LCM material.
2. Prepare lysates within 1 week of microarray printing. Long-term storage (>6 months) of protein lysates may result in protein degradation or diminished protein yield. If needed, store protein lysates at –80 °C or in the vapor phase of liquid nitrogen.

### 3.1.1. Preparation of Cell Culture Lysate

1. Prepare cell culture lysis buffer:

   > Cell Culture Lysis Buffer (per ml)
   > 915 µl TPER™
   > 60 µl 5 M NaCl (300 mM final conc.)
   > 10 µl 100 mM orthovanadate (1 mM final conc.) (boil vanadate 10 min in $H_2O$ to solubilize)
   > 10 µl 200 mM PEFABLOC (AEBSF)
   > 5 µl 1 mg/ml Aprotinin (5 µg/ml final conc.)
   > 1 µl 5 mg/ml Pepstatin A (5 µg/ml final conc.)
   > 1 µl 5 mg/ml Leupeptin (5 µg/ml final conc.)
   > Note: final concentration of NaCl is 450 mM because TPER™ reagent contains 150 mM NaCl.

2. Thaw cell pellets on ice for 20 min to 1 h depending on size of pellet.
3. Spin at approximately $900 \times g$ for 10 min in refrigerated (4 °C) centrifuge and remove any excess liquid.
4. If cells were stored in DMSO, wash cell pellet twice with PBS without calcium or magnesium. Wash in approximately 15 ml of PBS without calcium or magnesium, spin at approximately $900 \times g$ for 10 min. Decant supernatant and repeat wash step with an additional 15 ml of PBS without calcium or magnesium. Thoroughly decant supernatant.
5. Suspend cell pellet in appropriate volume of lysis buffer, vortex 15 s, spin briefly, and incubate on ice for 20 min. Volume of lysis buffer: $1 \times 10^6$ cells/100 µl of extraction buffer or, for lymphocytes, $1 \times 10^6$ cells/µl of extraction buffer.
6. Spin at approximately $3000 \times g$ for 5 min and transfer supernatant to clean tube.
7. Perform protein assay of choice to determine protein concentration.
8. Dilute sample with 2× Tris-glycine SDS Sample Buffer (Invitrogen) +2.5% 2-mercaptoethanol to 2 mg/ml based on the results of the protein assay (*see* **Note 5**).
9. Store lysate at –80 °C if needed.

## *3.2. Preparation of Tissue Lysate*

The limited volume of the total cellular material from biopsy samples makes it necessary to use a slight modification of the lysis buffer recipe as compared with the cell culture lysis buffer. The following protocol describes a method for preparing whole cell lysates from LCM cells. The limited biopsy sample generally precludes the use of total protein assays before microarray printing. To compensate for this, one microarray slide is stained for total protein using a Sypro Ruby Protein blot stain (*see* **Subheading 3.3.3.**).

### *3.2.1. Prepare Tissue Lysis Buffer*

1. Pipette 50 µl 2-mercaptoethanol, 950 µl 2× Tris-glycine SDS-loading buffer and 1 ml TPER™ into a clean plastic tube. Mix well.
2. Add 10 µl Sigma Protease Inhibitor Cocktail for each 1 ml of lysis buffer.
3. If desired, add 10 µl of 1 M orthovanadate to each 1 ml of lysis buffer (final concentration 10 mM orthovanadate).
4. Prepare daily. Store at room temperature.
5. As a rule of thumb, use 15 µl of lysis buffer to solubilize15,000 cells for microdissected tissue. More than one LCM cap may be used with a given volume of lysis buffer to concentrate the amount of protein in a given volume of lysis buffer (*see* **Note 6**).
6. Pipette the desired volume of lysis buffer in the bottom of the 0.5-ml microcentrifuge tube. Take care to avoid leaving droplets of lysis buffer near the lip of the tube. Place the LCM cap snugly on the tube. Invert the tube and mix well. Do not vortex.
7. Place the tube, cap side down, in an oven or heat block at 70 ± 2 °C for 15 min.
8. Mix well after 15 min. Place back in oven for an additional 15 min. Mix well again.
9. Spin tube at approximately 3000 × *g* for 1–2 min. Remove and discard LCM cap (*see* **Note 7**).
10. Transfer extracted protein sample to a clean, labeled screw cap tube.
11. Boil samples for 5–10 min before storing/printing microarrays. Store at –80 °C if necessary.

### *3.2.2. Protein Tissue Lysate Microarray Printing*

Each microarray comprises multiple samples, printed in dilution curves (usually a series of serial 1:2 dilutions), on a single slide. This format allows the comparison of multiple samples across an array for a given antibody (*see* **Fig. 1**). Additionally, the dilution curve allows each antibody affinity to be matched with its optimal sample concentration and ensures that every analyte is analyzed and compared in its linear dynamic range. Nitrocellulose cannot be effectively stripped and reprobed; therefore, each slide is probed with a

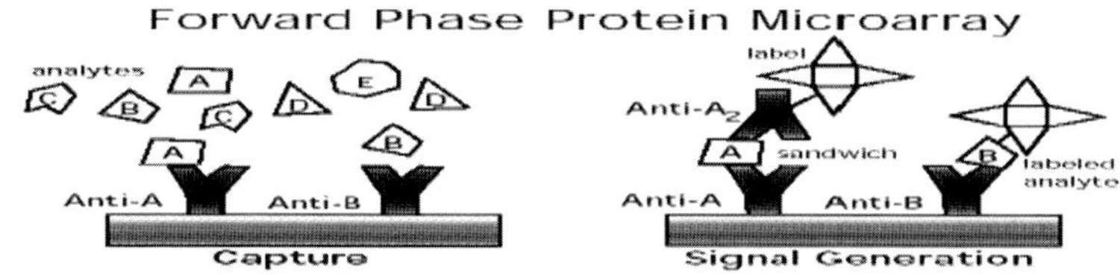

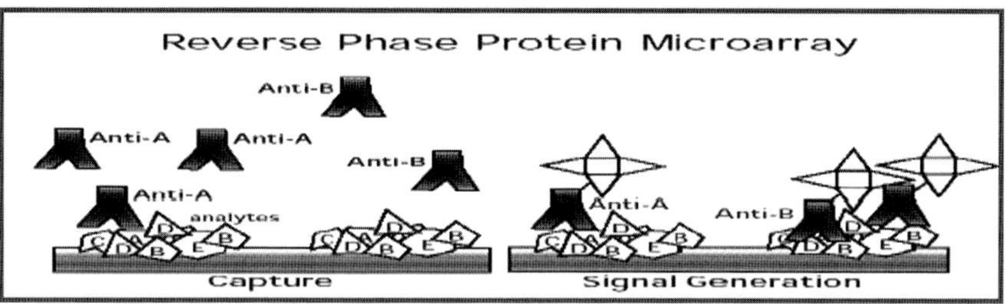

Fig. 1. Typical format for Forward (top) and Reverse (bottom) phase protein microarrays. The reverse phase format employs spots that comprise the analyte itself, usually in the form of a whole cell lysate or body fluid. An important advantage of the reverse phase array is the requirement for only one antibody, which greatly expands the number of analytes that can be measured in a given extract.

different antibody, generating a set of microarray slides for each set of antibody probes.

Quality control and comparison of microarrays across platforms and time requires the use of control lysates and calibrators that are printed on each array. These samples may be test samples, commercial cell lysates, peptides, or peptide mixtures. Ideally, the control lysate should contain the proteins(s) of interest that are being investigated in the test samples in a limited series of overall concentrations. For example, a series of controls could be an analyte at three predetermined concentrations of low, medium, and high levels. Control samples should be diluted in the same manner as the test samples, and the same control lysate should be printed on each array. Calibrators, such as a phosphopeptide, are arranged in a longer dilution curve such that the dynamic range of the assay is represented within the curve, providing a facile means of determining the level of any analyte in any experimental sample on the slide (*see* **Fig. 2**). The controls and calibrators allow for effective bridging of a given analyte across time and arrays.

1. A minimum volume of 30 µl of lysate is required for printing microarrays in a dilution format from a 384-well microtiter plate. Fifteen microliter of neat sample is used to prepare serial 1:2 dilutions of the lysate. An additional 15 µl of lysate is required for the neat sample.

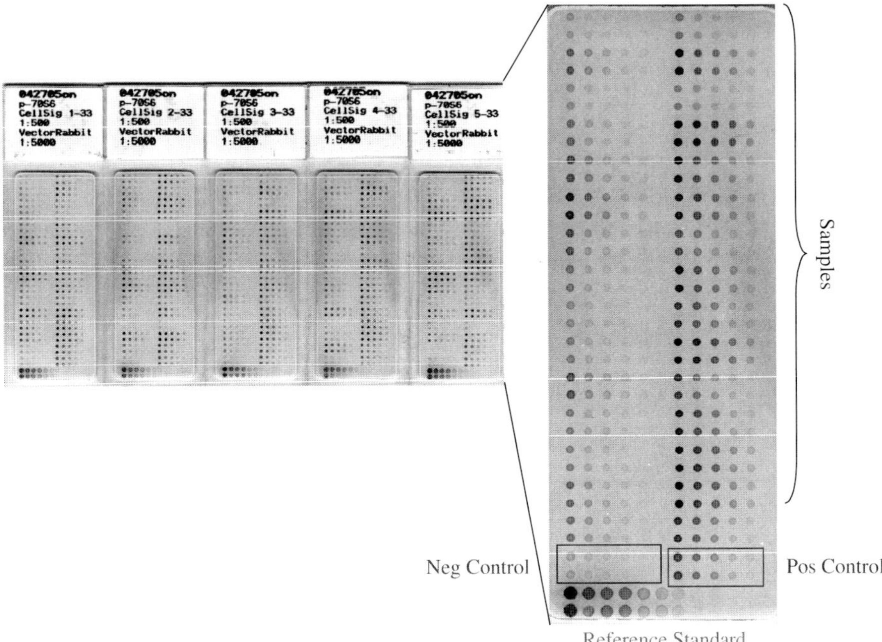



Fig. 2. Typical reverse phase protein microarrays (RPA) layout with controls and calibrators. A series of experimental test samples are arrayed in a series of five 1:2 dilutions (from left to right). Each sample is printed in duplicate, and the last spot in each series is a buffer alone control. On the bottom of every slide, a positive and negative control is printed as well as a reference standard calibrator.

2. A minimum volume of 50 µl of lysate is required for printing microarrays in a dilution format from a 96-well plate (25 µl of lysate to prepare the dilutions and 25 µl for the neat lysate).

3. Prepare additionallysis buffer for diluting the samples in a microtiter plate: 50 µl 2-mercaptoethanol, 950 µl 2× Tris-glycine SDS-loading buffer and 1 ml TPER™. Mix well. Prepare daily. Store at room temperature. It is not necessary to add additional protease or phosphatase inhibitors.

4. Boil whole cell lysate: place screw cap tube containing the thawed lysate sample in a boiling water bath or heat block at 100 °C for 5 min. Remove tubes from heat and allow tubes to cool to room temperature. Spin briefly in a microcentrifuge.

5. Dilute samples to be printed on the array in appropriate dilutions in a microtiter plate. Typical dilutions are 1:2 serial dilutions with the final dilution lysis buffer only (neat, 1:2, 1:4, 1:8, 1:16, buffer only). The buffer-only serves as a negative control spot.

6. Set-up the Aushon 2470 arrayer:

   1. Select microtiter plate—96 well or 384 well.
   2. Select number of microtiter plates.
   3. Select number of hits per spot. Five hits/spot is recommended with a 384-well plate. Ten hits/spot is recommended with a 96-well plate (*see* **Note 8**).
   4. Position nitrocellulose slides on printing platen. Load microtiter plate. Turn on humdifier if desired to achieve 30–70% humidity.
   5. Program appropriate dot spacing based on number of samples, replicate spots, and size of nitrocellulose pad (*see* **Note 9**).
   6. Print microarrays. Store printed slides in a slide box, inside a ziplock-style plastic bag. Add desiccant to the plastic bag before freezing. Store printed microarrays at –20 °C.

### 3.3. Immunostaining

#### 3.3.1. Microarray Slide Pre-Treatment and Blocking

Comparison of multiple phosphoproteomic endpoints for a group of samples on the array permits evaluation of the interconnected cell-signaling proteins in the sample populations.

Each microarray is probed with a single primary antibody directed against the protein of interest. Microarray slides used for immunostaining should be blocked before staining. Microarray slides used for Sypro Ruby protein staining *do not* require blocking.

1. Remove microarray slide(s) from freezer and leave at room temperature for approximately 5–10 min.
2. Prepare a 1× solution of Reblot™ mild solution in deionized water. Wash microarray slides with gentle rocking in 1× Reblot™ mild solution for 15 min (*see* **Note 10**).
3. Discard Reblot™ solution. Wash slides in PBS without calcium or magnesium twice, for 5 min each.
4. Block slides in I-block™ solution for a minimum of 1 h at room temperature (*see* **Note 11**). Preparation of I-block™:
5. Pour 500 ml of 1× PBS without calcium and magnesium into a 1000-ml beaker and add a magnetic stir bar.
6. Add 1 g of I-Block™ powder. Place beaker on hot plate with magnetic stirrer. Heat gently with stirring until solution becomes clearer (∼5–15 min). Do not boil.
7. Allow I-Block™ solution to cool to room temperature.
8. Add 500 μl of Tween20.
9. Store at 2–4 °C for up to 7 days.

### 3.3.2. Microarray Immunostaining

Immunostaining requires the use of control slides for determining background staining due to secondary antibody alone. Each species-specific secondary antibody should be used to stain an individual microarray. Therefore, for any immunostaining run, one slide must be used as a negative (secondary antibody alone) control slide. The total number of slides to be stained is determined by the number and species of primary antibodies. For example, if three anti-rabbit primary antibodies and one mouse primary antibody are selected, a total of 6 microarray slides will be needed (4 antibodies + 2 controls).

1. Select primary antibodies of choice.
2. Select species specific, biotinylated secondary antibodies.
3. Prepare CSA reagents following manufacturer's directions (*see* **Note 12**).
4. Prepare 1X Tris-buffered saline Tween20 (TBST) buffer per manufacturer's directions. Fill water reservoir with deionized water. Empty waste container if appropriate.
5. Load reagents and microarray slides on the Autostainer (*see* **Note 13**).
6. Prime water. Prime buffer. Start run.
7. Remove slides promptly at end of Autostainer run. Do not allow the Autostainer to add TBST to the slides after staining is complete (*see* **Note 14**). Allow slides to air dry away from direct light. Covering the slides lightly with a paper towel is adequate to reduce exposure to direct light.
8. Label slides as to which antibody was used for immunostaining.
9. Scan slides and store the stained slides in the dark at room temp.

### 3.3.3. Sypro Ruby Total Protein Staining of Microarray

The limited sample material from microdissected biopsy lysates precludes the use of spectrophotometeric analysis of total protein prior to microarray printing. A microarray slide stained for total protein serves as a tool for normalizing spot intensity between samples (*see* **Note 15**).

1. Remove microarray slide(s) from freezer and leave at room temperature for approximately 5–10 min (*see* **Note 16**).
2. Prepare fixative solution: 5.0 ml methanol, 3.5 ml acetic acid, and 41.5 ml dH$_2$O (final concentrations 10% methanol, 7% acetic acid). Mix well and store tightly sealed.
3. Wash slide in dH$_2$O 2× for 5 min each with continuous agitation or rocking.
4. Fix slides by immersing in fixative solution (volume sufficient to cover slides) for 15 min. Place on orbital shaker or rocker.
5. Wash in dH$_2$O 4× for 5 min each with continuous agitation or rocking.
6. Immerse slides in Ruby Red stain (volume sufficient to cover slides) for 30 min at room temperature, in the dark. Cover container with aluminum foil to prevent photobleaching of the stain. Place on orbital shaker or rocker.

7. Wash in dH$_2$O 4× for 1 min each, in the dark, with continuous agitation or rocking.
8. Air dry slides at room temperature in the dark.
9. Scan slides with a standard UV transilluminator or a laser scanner. Refer to http://probes.invitrogen.com/media/pis/x11791.pdf for a list of compatible scanning platforms.

## 3.4. Scanning and Data Analysis

The chromogenic detection system described allows the microarrays to be scanned on any high resolution scanner [1200 dots per inch (dpi)] with software capable of a 14- or 16-bit scanning option for grayscale. Electronic images may be imported into a variety of spot analysis software programs, such as PSCAN (http://abs.cit.nih.gov), MicroVigene™ (Vigene Tech), or ImageQuant® (Amersham Biosciences). Each array is scanned, the spot intensity analyzed, data are normalized to total protein (*see* **Note 17**), and a standardized, single data value is generated for each microarray sample (*see* **Fig. 3**). The data may be used to generate histograms, dendograms based on

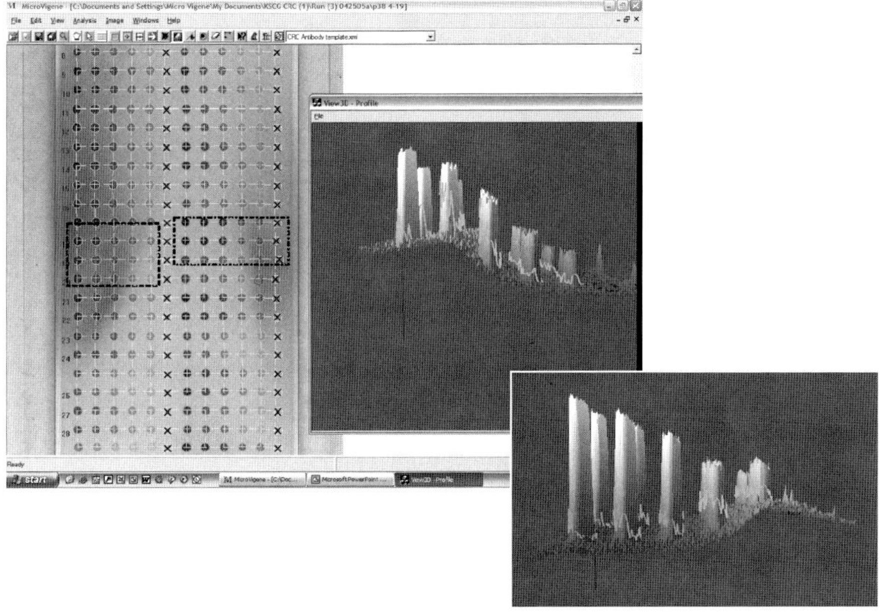

Fig. 3. Data analysis performed by using MicroVigene (VigeneTech, Billerica, MA, USA) software program. Spot finding, regional and local background subtraction, and linear dynamic range finding are performed by the software program. Combined with the calibrator, this method provides for a reproducible and quantifiable value for every sample.

hierarchical clustering algorithms, or Bayesian clustering analysis for generation of cell-signaling network profiles.

### 3.4.1. Scanner Settings

1. Place the microarray slides on the scanner.
2. Scan at 600–1200 dpi. Convert the image to grayscale. Save image as a 14- or 16-bit grayscale image.
3. Save the image as a tiff file, compatible with the spot analysis software.

## 4. Notes

1. Methods are under development for analysis of protein from formalin-fixed tissues, although protein yield may be less than that from frozen tissue. Extraction of high-yield, quality protein from formalin-fixed tissue is generally difficult because of the extensive cross-links formed in formalin-fixed samples. Currently, the optimal tissue specimen for use with protein lysate microarrays is fresh tissue, immediately embedded in a cyroprotectant solution and frozen or snap-frozen tissue.
2. Reblot™ solution is manufactured in two strengths: mild and strong. It is imperative that the mild solution be used with nitrocellulose-coated slides. The Reblot™ strong solution may cause the nitrocellulose to detach from the glass slide.
3. The primary antibody should be validated by western blot before use on a microarray. A validated antibody should show a single band at the specified molecular weight. Primary antibodies should be unconjugated and may be any species, with the only caveat being the biotinylated secondary antibody should be species matched with the primary antibody. Primary and secondary antibody concentration should be optimized for use on a microarray before use with patient samples. Typically, primary antibody dilutions of 1:250, 1:500, and 1:1000 are used to immunostain a set of microarrays. The secondary antibody concentration is held constant. Comparison of signal to noise ratio is used to determine the optimal concentration to use on the microarray. Secondary antibodies are optimized in a similar manner. A set of microarrays is immunostained with secondary antibody alone at a variety of concentrations and antibody diluent is used as the primary antibody. In general, the secondary antibody dilution that shows minimal staining is the optimal dilution. Verification of the optimal secondary antibody concentration may be performed by staining a microarray with an optimized primary antibody and the secondary antibody dilution of choice and accessing the signal to noise ratio.
4. Sypro Ruby blot stain is a ruthenium complex, exhibiting luminescence upon excitation with either UV-B or blue light. The luminescence may be visualized with UV epi-illumination sources, UV or blue-light transilluminators, or laser-scanning instruments with excitation light at 450, 473, 488, or 532 nm. The emission peak of the ruthenium complex is approximately 618 nm *(24)*.
5. The binding capacity of the nitrocellulose is 9 μg/mm$^3$ with a 0.2-μm pore membrane *(25)*. Samples with protein concentrations greater than 2 mg/ml may

saturate the nitrocellulose with concomitant loss of the dilution series in the printed diluted spots.

6. A common issue with microdissected tissue is minimal cell number on any given LCM cap. To effectively increase the cell number and protein concentration for a volume of lysis buffer, it may be necessary to solubilize 2–3 LCM caps in 1 vol of lysis buffer. This is done by sequentially solubilizing the cells on LCM caps in the same volume of lysis buffer. To illustrate the process, one LCM cap is solubilized in lysis buffer. This first cap is discarded and a fresh LCM cap containing microdissected material is placed on a tube containing the lysis buffer used previously to solubilize the first LCM cap. The cells on this second cap are solubilized in the lysis buffer, effectively increasing the protein concentration in a given volume of buffer.

7. After solubilizing cells from an LCM cap, the efficiency of extraction may be determined by examining the cap surface with the aid of a microscope. Place the cap, polymer side down, on a clean glass microscope slide. Observe the surface of the polymer with a standard light microscope. Cellular material should not be visible on the polymer surface. If cells are still present, place the cap on the tube containing lysis buffer and continue heating the cap at 70 °C for an additional 15 min. Repeat the above process until the cells are solubilized.

8. The Aushon Biosystems 2470 Arrayer can effectively print multiple depositions of the same sample on the exact same spot, thus effectively increasing the protein concentration/spot. Printing 5–10 hits/spot maximizes the protein-binding capacity of nitrocellulose without saturating the nitrocellulose. The recommendation to print 10 hits/spot from a 96-well plate and 5 hits from a 384-well plate is based on dot-spacing restrictions in the $x$- and $y$-axes from the different microtiter plate formats.

9. Typical microarrays with human samples are printed in triplicate for quality control and statistical analysis. Dot spacing is a function of pin diameter— dot spacing must be a minimum of 1.5× the pin diameter. FAST™ slides are manufactured in a variety of formats, with a 20 × 50 mm nitrocellulose pad being compatible with reverse phase microarray format and the DakoCytomation Autostainer.

10. Wash time is critical with the Reblot™ mild solution. Extended wash times may cause the nitrocellulose to detach from the glass.

11. Microarrays may be blocked overnight at 4 °C. The minimum blocking time is 1 h.

12. The Autostainer grid layout is a modification of the CSA grid provided by the manufacturer. Additional TBST rinse steps have been added to ensure adequate rinsing of the microarray slides. About 450 µl of reagent is added to the slide in each of the three drop zones.

13. Do not let the microarray slides dry while loading and before starting the Autostainer run. If necessary, pour I-block or 1× TBST on the slides. If the ambient humidity is low, a shallow tray of water may be placed inside the Autostainer during the staining run. Operate the Autostainer with the lid closed.

14. The DakoCytomation Autostainer is designed primarily for immunohistochemistry. In these procedures, the slides are kept moist with periodic rinses of TBST. The high salt content causes crystal formation on the nitrocellulose if the TBST is allowed to dry on the microarray slide. If crystal formation inadvertently happens, it may be possible to dissolve the crystals by washing the microarrays with PBS + 0.2% Tween20 for 1 h and then washing with PBS only. A final rinse in water is usually sufficient to remove the crystals. Additionally, the Autostainer may be programmed to include a final water step. Adding an "auxiliary" reagent step to the end of the program allows the Autostainer to run in an "overnight" mode. This auxiliary reagent is deionized water and the time is 840 min (14 h). The microarray slides will be rinsed with water following the DAB detection step. The Autostainer will remain idle, waiting to dispense water in 840 min, allowing the operator to return in 14 h.

15. Syrpo Ruby staining is a permanent protein stain, detected by using fluorescence with an excitation wavelength of 300 or 480 nm and an emission wavelength of 618 nm. The stain is photostable, allowing long emission lifetime and the ability to measure fluorescence over a longer time frame, minimizing background fluorescence.

16. Do not block (with I-block) the microarray slide used for total protein staining. Because of slight changes in the protein concentration per spot from the first slide printed to the last slide printed, an average protein concentration per spot for all slides printed within a run may be determined from one slide. Typically, a slide representing the median slide of the printing run is selected for total protein staining.

17. For example, if 25 slides were printed in a run, slide 12 would be selected for total protein staining. Dot spacing on the microarray is a critical factor for ensuring successful spot intensity analysis. Careful planning of the array configuration before printing is essential. Spots placed too closely prevent discrimination of background areas from spot area with most spot analysis software programs. The 384-well format results in 12 horizontal spots, in two dilution curves for two different samples.

## References


1. Liotta L. A., Espina V., Mehta A. I., et al. (2003) Protein microarrays: Meeting analytical challenges for clinical applications. *Cancer Cell* 3(**4**), 317–25.
2. Haab B. B., Dunham M. J., and Brown P. O. (2001) Protein microarrays for highly parallel detection and quantitation of specific proteins and antibodies in complex solutions. *Genome Biol* 2(**2**), RESEARCH0004.
3. Macbeath G. and Schreiber S. L. (2000) Printing proteins as microarrays for high-throughput function determination. *Science* 289(**5485**), 1760–3.
4. Macbeath G. (2002) Protein microarrays and proteomics. *Nat Genet* 32 (Suppl.) 526–32.

5. Paweletz C. P., Charboneau L., Bichsel V. E., et al. (2001) Reverse phase protein microarrays which capture disease progression show activation of pro-survival pathways at the cancer invasion front. *Oncogene.* 20(**16**), 1981–9.

6. Zhu H. and Snyder M. (2003) Protein chip technology. *Curr Opin Chem Biol* 7(**1**), 55–63.

7. Wilson D. S. and Nock S. (2003) Recent developments in protein microarray technology. *Angew Chem Int Ed Engl* 42(**5**), 494–500.

8. Templin M. F., Stoll D., Schrenk M., et al. (2002) Protein microarray technology. *Trends Biotechnol* 20(**4**), 160–6.

9. Schaeferling M., Schiller S., Paul H., et al. (2002) Application of self-assembly techniques in the design of biocompatible protein microarray surfaces. *Electrophoresis* 23(**18**), 3097–105.

10. Weng S., Gu K., Hammond P. W., et al. (2002) Generating addressable protein microarrays with PROfusion covalent mRNA-protein fusion technology. *Proteomics* 2(**1**), 48–57.

11. Petach H. and Gold L. (2002) Dimensionality is the issue: use of photoaptamers in protein microarrays. *Curr Opin Biotechnol* 13, 309–314.

12. Lal S. P., Christopherson R. I., and Dos Remedios C. G. (2002) Antibody arrays: An embryonic but rapidly growing technology. *Drug Discov Today* 7(**Suppl. 18**), S143–9.

13. Humphery-Smith I., Wischerhoff E., and Hashimoto R. (2002) Protein arrays for assessment of target selectivity. *Drug Discovery World* 4(**1**), 17–27.

14. Bobrow M. N., Harris T. D., Shaughnessy K. J., and Litt G. J. (1989) Catalyzed reporter deposition, a novel method of signal amplification. Application to immunoassays. *J Immunol Methods* 125(**1–2**), 279–85.

15. Bobrow M. N., Shaughnessy K. J., and Litt G. J. (1991) Catalyzed reporter deposition, a novel method of signal amplification. II. Application to membrane immunoassays. *J Immunol Methods* 137(**1**), 103–12.

16. Hunyady B., Krempels K., Harta G., and Mezey E. (1996) Immunohistochemical signal amplification by catalyzed reporter deposition and its application in double immunostaining. *J Histochem Cytochem* 44(**12**), 1353–62.

17. King G., Payne S., Walker F., and Murray G. I. (1997) A highly sensitive detection method for immunohistochemistry using biotinylated tyramine. *J Pathol* 183(**2**), 237–41.

18. Petricoin E., Wulfkuhle J., Espina V., and Liotta L. A. (2004) Clinical proteomics: Revolutionizing disease detection and patient tailoring therapy. *J Proteome Res* 3(**2**), 209–17.

19. Grubb R. L., Calvert V. S., Wulkuhle J. D., et al. (2003) Signal pathway profiling of prostate cancer using reverse phase protein arrays. *Proteomics* 3(**11**), 2142–6.

20. Wulfkuhle J. D., Aquino J. A., Calvert V. S., et al. (2003) Signal pathway profiling of ovarian cancer from human tissue specimens using reverse-phase protein microarrays. *Proteomics.* 3(**11**), 2085–90.

21. Liotta L. A., Kohn E. C., and Petricoin E. F. (2001) Clinical proteomics: Personalized molecular medicine. *JAMA* 286(**18**), 2211–4.

22. Petricoin E. F., Zoon K. C., Kohn E. C., Barrett J. C., and Liotta L. A. (2002) Clinical proteomics: translating benchside promise into bedside reality. *Nat Rev Drug Discov.* 1(**9**), 683–95.

23. Wulfkuhle J. D., Edmiston K. H., Liotta L. A., and Petricoin E. F. (2006) Technology Insight: Pharmacoproteomics for cancer-promises of patient-tailored medicine using protein microarrays. *Nat Clin Pract Oncol* 3(**5**):256–68.

24. Berggren K., Steinberg T. H., Lauber W. M., et al. (1999) A luminescent ruthenium complex for ultrasensitive detection of proteins immobilized on membrane supports. *Anal Biochem* 276(**2**), 129–43.

25. Tonkinson J. L. and Stillman B. A. (2002) Nitrocellulose: A tried and true polymer finds utility as a post-genomic substrate. *Front Biosci* 7, c1–12.

# 9

# Antibody Arrays for Determination of Relative Protein Abundances

## Grigoriy S. Chaga


## Summary

As a large number of genome-sequencing projects reached completion, the attention of the scientific community is turning toward understanding the structure-functions of gene translation products—the proteins as well as the complete complement of proteins—the proteome. One goal of proteomics is to correlate changes in protein abundance with biological processes and disease states. To help accelerate this avenue of proteomics, a significant effort has been devoted to the development of multiplexed methods for protein analyses. We have developed an Antibody Microarray, a chip-based technology for multiparallel determination of relative abundance of hundreds of proteins. The Antibody Microarray is composed of hundreds of distinct monoclonal antibodies printed at high density on a glass slide. It utilizes a novel experimental setup and data analysis algorithm, which enables scientists to assay hundreds of cytosolic, nuclear, and membrane-bound proteins with a single experiment. Examples of biological samples that are analyzed on the Antibody Microarray include tissue samples, cell cultures, and body fluids.

**Key Words:** Antibody arrays; protein arrays; protein abundance; multi-parallel protein profiling.


## 1. Introduction

The realignment of focus of the scientific community on protein function has generated a strong need for novel, high-throughput analysis tools and systems that can determine the abundance, post-translational modifications, and biological function of the proteome encoded by the human genome.

Extensive development in the field of DNA arrays for high-throughput analyses of gene expression levels for more than a decade has resulted in the


From: *Methods in Molecular Biology, vol. 441: Tissue Proteomics: Pathways, Biomarkers, and Drug Discovery*
Edited by: B.C.-S. Liu © Humana Press, Totowa, NJ

ability to generate a high volume of data on a number of species and tissue types.

However, it is becoming more and more apparent that there is low correlation between mRNA levels and protein abundance *(1,2)*, and, as a result, the development of tools for multiplex protein abundance determination is becoming increasingly important.

Number of publications cover the topic of the application of antibodies as capture and/or detection reagents in high-throughput analyses of protein abundance *(3–11)*.

Our antibody microarray is designed to enable scientists to compare the abundance of 507 proteins from a wide type of biological samples.

We utilize an experimental setup and mathematical algorithm for determination of relative protein abundance from directly labeled native protein samples applied to an array of antibodies. The application of the proposed experimental system compensates internally at each array element for a number of deficiencies in array experiments such as differential labeling efficiency in dual color assay systems, differential solubility of protein molecules in dual color assay systems, and differential affinity of capture reagents toward proteins labeled with two different fluorescent dyes. This system offers full compensation for variable amounts of capture reagents on separate array structures, as well as limited compensation for non-specific interactions between capture reagents and analytes. The proposed experimental strategy enables the use of a large number of capture reagents to develop a true multiplex analysis system that would yield complete relative protein abundance information in two biological systems.

Examples of the antibody microarray's utility include the studies of protein abundances in brain research *(12)*, apoptosis *(13,14)*, cancer research *(15,16)* as well as diabetes *(17)*.

What are some important requirements for a profiling antibody array system:

## 1.1. Sample Quality

It is difficult to overestimate the importance of the sample quality in multiplex analyses. Taking into account that antibody array samples are several orders of magnitude more complex than DNA array samples, the requirements are even more stringent. Important considerations are efficiency of extraction, true representation of the proteome inside the sample after the extraction and labeling, reproducibility of extraction and labeling, minimal loss of proteins during labeling, and removal of unincorporated dye. Our extraction and labeling system is designed to deal as well as possible with these requirements. Important features of the system are

- The extraction results in a sample that contains native proteins. Furthermore, it is not expected to result in protein–protein complex destabilization—additional protection of protein epitopes from over-labeling because of partial protection of the protein molecules surfaces that are involved in protein–protein interaction.
- The same buffer in which the extraction is carried out is used for labeling—i.e., there is no need for buffer exchange between the extraction and labeling steps.
- The extraction efficiency of the buffer has been determined to be more than 95% of that of sodium dodecyl sulfate (SDS)-boiling method. Our western analyses of marker proteins from various cellular compartments point to excellent representation of proteins from all cellular compartments (in some cases, as with cytochrome C it is better than the SDS-boiling method)—(*see* **Fig. 1**)

## 1.2. Experimental Setup

There are number of issues that arise when applying multiplex analyses to directly labeled samples:

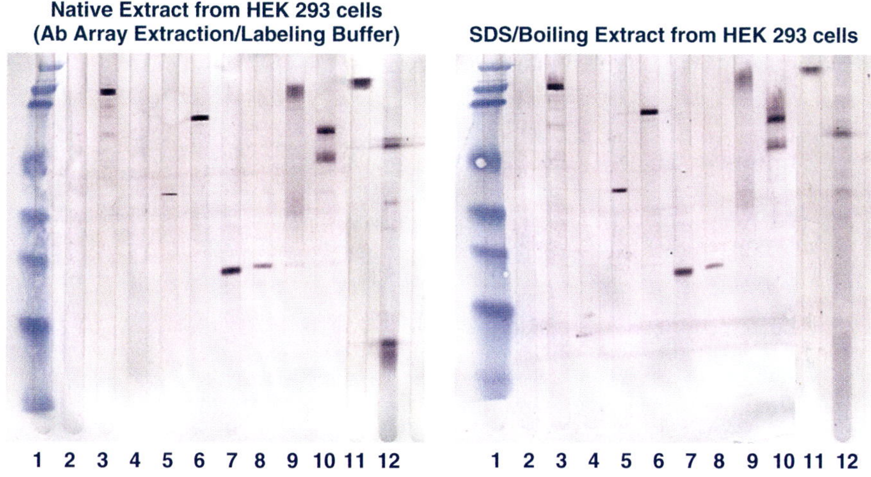


Fig. 1. Western analyses of extraction efficiency of the antibody microarray extraction/labeling buffer. Equal amounts of sample extracted by the extraction/labeling buffer and by sodium dodecyl sulfate (SDS)-boiling method (1% SDS in TST buffer for 5 min at 100°C) were separated by using SDS electrophoresis in twelve lanes (total of 20 µg per lane) each and each lane for both extracts was probed after a transfer of the proteins to a Polyvinylidene Difluoride (PVDF) membrane with a cell compartment-specific marker antibody. Lanes are 1. MW Standards, 2. Empty, 3. Vinculin—Cytoskeleton, 4. Empty, 5. ERK 1—Cytosol, 6. BIP/GRP78—Endoplasmic reticulum, 7. RAB 11—Endosomes, 8. RAB 5—Endosomes, 9. Lamp-1—Lysosome, 10. Nucleoporin p62—Nucleus, 11. Integrin b1—Plasma membrane, 12. Cytochrome C—Mitochondria.

- Lot-to-lot and even tube-to-tube variation in the amount of active dye for labeling of the sample.
- Differential labeling efficiency of the Cy5 versus Cy3 dye for the particular antigen analyte.
- Differential solubility of the analyte labeled with Cy5 versus that of the same analyte labeled with 3 dye.
- Differential affinity of the capture antibody toward the analyte labeled with Cy5 versus the analyte labeled with Cy3 dye.

### *1.3. Internally Normalized Ratio*

To address all the issues described above, we have developed an experimental setup and algorithm that compensates completely or significantly minimizes the impact of these factors. The experimental setup includes labeling each sample with both dyes at the same time and utilizing the same tube of label dye for both samples. In addition, we run dual color/reverse color incubation scheme—i.e., Sample A labeled with Dye 1 is mixed with Sample B labeled with Dye 2 and applied to Slide 1, whereas Sample B labeled with Dye 1 is mixed with Sample A labeled with Dye 2 and applied to Slide 2 (*see* **Fig. 2**)

The internally normalized ratio (INR) from any two samples (e.g., samples A and B) can be derived as follows:

The ratio of the ratios from both slides can be defined as

$$\frac{R_1}{R_2} = \frac{A^{Cy5}}{B^{Cy3}} \Big/ \frac{B^{Cy5}}{A^{Cy3}} = \frac{A^{Cy5}}{B^{Cy3}} \times \frac{A^{Cy3}}{B^{Cy5}} = X \times Y \tag{1}$$

where $R_1$ is the Cy5/Cy3 ratio obtained from Slide 1, and $R_2$ is the Cy5/Cy3 ratio obtained from Slide 2.

In the ideal case, when the amount of any particular antigen is equal in both samples and the effect of differential labeling, solubility, affinity, etc., is the same for the antigen labeled with both labels (i.e. $X = Y$), INR can be defined as

$$INR = \sqrt{X^2} = \sqrt{R_1 / R_2} \tag{2}$$

The application of INR algorithm in conjunction with our experimental setup results in significant decrease of false-positive results as exemplified in **Fig. 1**:

Total protein from HeLa cells was extracted and labeled as described in Section 3. Methods, 3.1. Large Scale Protocol. Two glass slides were incubated with a Cy5 and Cy3 mix at 10 µg of protein per channel. The resulting Cy5/Cy3 ratios were plotted in **Fig. 3a**, and the INR obtained from the two slides was plotted in **Fig. 3b**. There is a significant number of false-positive data displayed

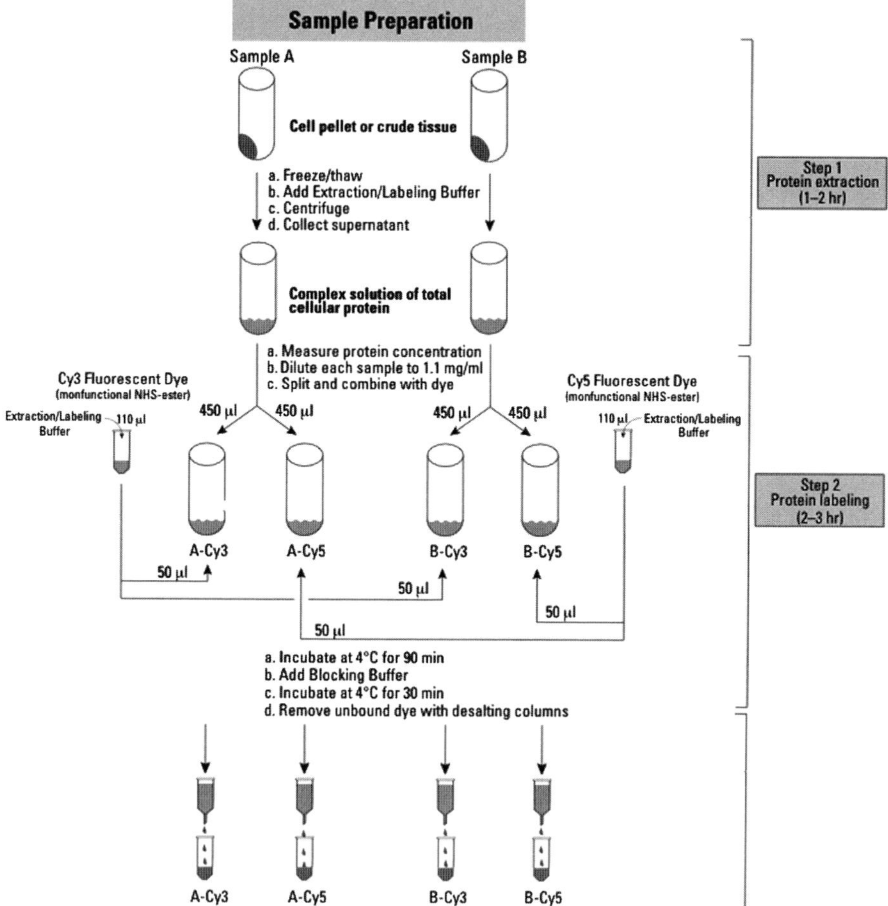



Fig. 2. Schematic presentation of the experimental setup used in the antibody microarrays.

in **Fig. 3a**—i.e., more than 2:1 ratio detected. The application of INR to the ratios from the both slides compensates completely—all reported data points are between 0.7 and 1.3.

By applying the INR algorithm, we successfully compensate for all four factors discussed above that contribute to generation of false-positive data (*see* **Fig. 3b**).

An additional factor that affects the quality of the generated data is the differences in the amount of immobilized capture antibody between arrays.

This issue is also addressed by using the dual color/reversed color experimental setup. Significant differences in the amount of the active capture antibody between the arrays would eventually result in inability to compensate for it even by this experimental design, but they have to be really drastic—more

(a)

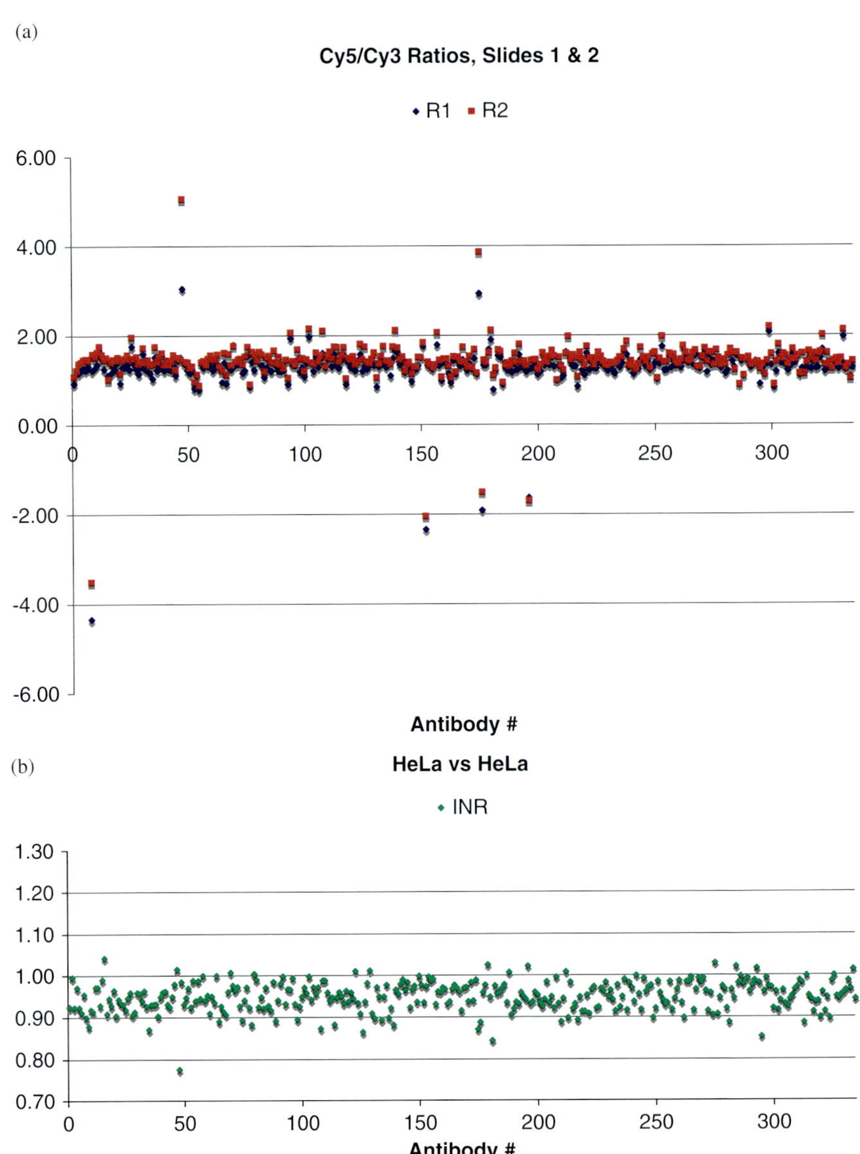



(b)

Fig. 3. Effect of internally normalized ratio (INR) on the quality of generated data. a) Data obtained from identical experiments with single slides utilizing dual color on the same sample. b) Data obtained from the application of the INR algorithm on the results from Fig. 3a.

than 5:1 or even 10:1, as at lower ratios the type of the assay (competitive assay) would report similar values for the ratio of the analyte in the two samples.

## 1.4. General Considerations

- Wear laboratory gloves whenever handling Ab microarrays. Alternatively, use tweezers to manipulate slides.
- Always hold slides at the end nearest the affixed data label. (Note: This label includes unique identifying information for the array.)

## 1.5. Orienting the Microarrays

- To assist you in identifying the different blocks of the printed area after scanning, spots of Cy3/Cy5-labeled bovine serum albumin (BSA) have been printed on the slide. These BSA spots serve as orientation markers that you can use to align the grid of your array analysis software. The Cy3/Cy5-labeled spots are located at or near the outermost corners of the printed area except the right bottom corner where is a negative control. The precise coordinates are given on the Product Analysis Certificate included with your microarrays.
- Note that the data label is affixed to the printed surface of the slide.
- Please see the enclosed Product Analysis Certificate for the identity and location of all array spots.

## 1.6. Using the Storage Chamber

- Ab microarrays are supplied inside a green-capped storage vial. Do not remove the microarray slides until you have labeled your protein samples and are ready to start the incubation.
- An empty, green-capped storage vial is also supplied. Use this vial to dry the microarray slides.

## 1.7. Choosing Between Large-Scale and Small-Scale Protocols

In completing an Ab microarray analysis, you have the option of using either a Large-Scale or Small-Scale Protein Extraction & Labeling protocol. The key differences between these two protocols are summarized in **Table 1**. Both protocols yield sufficient protein to perform a single antibody microarray analysis. The large-scale method, however, yields more than enough protein (*see* **Note 1**). If desired, you can use the additional protein for other types of analyses performed in conjunction with the antibody microarray procedure. For example, portions of the remaining fluorescently labeled protein could be used for western blotting or two-dimensional gel electrophoresis. Note, however, that we do not recommend storing labeled protein for long periods because of the potential for protein degradation.

**Table 1**
**Comparison of Extraction/Labeling Scales**

|                                          | Small-scale | Large-scale |
|------------------------------------------|-------------|-------------|
| Quantity of tissue or cells              | 15–25 mg    | 50–100 mg   |
| Normal protein yield                     | 250–500 μg  | 2–3 mg      |
| Amount needed for the labeling step      | 200 μg      | 1 mg        |

## 2. Materials

### 2.1. Samples for Extraction of Proteins

Note: The amount of sample is primarily a function of the size of the dye aliquots provided commercially (require 1 mg of protein per labeling reaction) rather than the amount used in the array experiments.

1. Large-scale—between 50 and 100 mg of tissue are necessary to perform an extraction that would result in 2.5–5 mg of total protein. As the labeling reaction is carried out with approximately 0.5 mg of protein for each fluorescent dye (i.e., ~1 mg of total protein), there is a sufficient amount for analyses, labeling, and downstream validation.
2. Small-scale—approximately 20 mg of tissue is necessary from tissue cultures to carry out the extraction, labeling, and the incubation per sample.
3. Alumina (Sigma, St. Louis, MO, USA Cat. No. A-2039; for disintegrating tissue samples).
4. Mortar and pestle (for grinding tissue).
5. BCA Protein Assay Reagent Kit (Pierce Biotechnology, Rockford, IL, USA; Cat. No. 23225 or 23227).
6. BSA (protein standard).

### 2.2. Labeling of Proteins

1. Cy5 mono-Reactive Dye Pack (Amersham Biosciences, GE Healthcare Bio-sciences Corp., Piscataway, NJ, USA; Cat. No. PA25001)
2. Cy3 mono-Reactive Dye Pack (Amersham Biosciences; Cat. No. PA23001)
3. 1.5- and 2.0-ml microcentrifuge tubes.

### 2.3. Removing Unbound Dye (Desalting)

1. 15- and 50-ml conical centrifuge tubes (e.g., BD Falcon TM conical centrifuge tubes)
2. Disposable PD-10 desalting columns (Amersham Biosciences; 17-0851-01)

These columns can be used for both Large-Scale and Small-Scale Protein Extraction & Labeling protocols.

## 2.4. Antibody Array Incubation

1. Clontech Ab Microarrays (Cat. No. 631790)—two arrays per kit
2. Clontech Ab Microarray Buffer Kit (Cat. No. 631786)—complete set of buffers for extraction, labeling, incubation, and washing of large and small scale experiments.
3. Rocking platform (to provide a constant "see-saw" motion during slide incubation and washing).
4. Swinging-bucket centrifuge (with adaptors for spinning 50-ml tubes).
5. Microcentrifuge.
6. Spectrometer capable of measuring absorbance at 552 and 650 nm.
7. Microarray slide scanner. You may use any scanner that is compatible with 75 × 25 × 1-mm slides and capable of dual-color analysis. The scanner must be capable of measuring Cy5 and Cy3 fluorescent labels.

## 2.5. Analysis of Results

1. Microsoft® Excel 97/98 (software application) Used for calculating INRs based on fluorescence data from a microarray analysis.
2. Clontech Ab Microarray Analysis Workbook and Axon Grid (Gal file) (available at http://bioinfo.clontech.com)

# 3. Methods

## 3.1. Large Scale Protocol

### 3.1.1. Extracting Protein from Tissue Samples

1. Before starting, chill the following items on ice or at 4°C:

   - Extraction/labeling buffer
   - One mortar and pestle
   - Two 2-ml microcentrifuge tubes
   - One 15-ml conical centrifuge tube

2. Transfer 50–100 mg of frozen tissue to a pre-chilled mortar (*see* **Note 2**).
3. Add 0.125–0.25 g of alumina to the mortar.
4. Use the pestle to grind the tissue until a paste is formed.
5. Add 1–2 ml of pre-chilled extraction/labeling buffer.
6. Mix the buffer into the paste using the pestle. When you finish, use a micropipette tip to scrape the paste that adheres to the pestle back into the mortar.
7. Transfer the extract to a pre-chilled 2-ml microcentrifuge tube.
8. While holding the pestle over the mortar, rinse the pestle with 1–2 ml of extraction/labeling buffer.
9. Combine the rinse with the original extract in a 2-ml tube. (Use a second 2-ml tube if the volume exceeds the tube's capacity.)
10. Centrifuge the suspension at 10,000 × *g* for 30 min.

11. While taking care not to disturb the pellet, transfer the supernatant to a pre-chilled 15-ml conical centrifuge tube.
12. Gently invert the tube to mix the lysate.
13. Measure protein concentration using Pierce's BCA Protein Assay Reagent Kit (*see* **Note 3**).
14. Dilute each sample to 1.1 mg protein/ml by adding the appropriate volume of extraction/labeling buffer. The final volume must be $\geq 1$ ml (*see* **Note 4**).
15. Proceed immediately with the next section.

### 3.1.2. Labeling of Proteins

Important: Complete **steps 1–10** rapidly without interruption. Once the Cy3 and Cy5 dyes are dissolved in buffer, they must be used immediately. The procedure below is for large-scale method.

1. Label four 1.5-ml microcentrifuge tubes: "A-Cy3," "A-Cy5," "B-Cy3," and "B-Cy5."
2. Dissolve the Cy3 dye in 110 µl of extraction/labeling buffer by adding the buffer directly to the tube in which the dye is supplied. Note: Each tube of dye contains a quantity of dye that will label approximately 1 mg of total protein.
3. Mix thoroughly by vortexing for 20 s.
4. Centrifuge the tube at moderate speed for 10 s to recover the liquid in the bottom of the tube.
5. Prepare a solution of Cy5 dye in the same manner by following **steps 2–4**.
6. Add 50 µl of Cy3 solution to tubes "A-Cy3" and "B-Cy3."
7. Add 50 µl of Cy5 solution to tubes "A-Cy5" and "B-Cy5."
8. Add 450 µl of protein sample A to tubes "A-Cy3" and "A-Cy5."
9. Add 450 µl of protein sample B to tubes "B-Cy3" and "B-Cy5."
10. Invert each tube three times to mix the contents. Then, centrifuge each tube at moderate speed for 10 s to recover the liquid in the bottom of the tube.
11. Incubate all four tubes on ice (or at 4°C) for 90 min. Mix each tube by inversion every 20 min.
12. Add 4 µl of blocking buffer to each tube.
13. Incubate each tube on ice (or at 4°C) for 30 min. Mix each tube by inversion every 10 min.

### 3.1.3. Removing Unbound Dye (Desalting) and Determination of Incorporated Dye

Use Amersham Biosciences PD-100 desalting columns to remove unbound dye as follows. (As long as you work quickly, **steps 1–7** can be performed at room temperature. Otherwise, if you have access to a cold room, we suggest you complete **steps 1–7** at 4°C.)

1. Set up and label four PD-10 desalting columns and four 2-ml microcentrifuge tubes: A-Cy3, A-Cy5, B-Cy3, and B-Cy5.

2. Prepare 100 ml of 1× desalting buffer by diluting 10× desalting buffer with the appropriate volume of Milli-Q-grade H2O. Store 1× desalting buffer in a clean plastic bottle. Note: Be sure that 1× desalting buffer has a pH = 7.4. Adjust the pH if necessary using dilute HCl or NaOH.

3. Equilibrate each column with 3 × 5 ml of 1× desalting buffer.

4. Apply the Cy3- and Cy5-labeled protein samples (~500 µl each) to the corresponding columns. Allow the protein sample to pass into the column.

5. Add 2 ml of 1× desalting buffer to each column. Allow the buffer to pass into the column to push the protein sample further along.

6. Place the 2-ml microcentrifuge tubes (from **step 1**) under the corresponding columns.

7. Elute each protein sample by applying 2 ml of 1× desalting buffer to each column. Collect the flowthrough.

8. Store the tubes on ice.

9. Measure protein concentration using Pierce's BCA Protein Assay Reagent Kit: Use 96-well microplate. Dilute samples 10:1 (20–200 µl with 180 µl of desalting buffer) into pre-labeled tubes (*see* **Notes 5–9**).

10. Estimate the average number of dye molecules covalently coupled to each protein. Follow the protocol in the Amersham Product Specification Sheet, but first see the instructions below for estimating the average number of coupled dye molecules (*see* **Notes 10** and **11**):

    - Measure Cy3 absorbance at 552 nm and Cy5 absorbance at 650 nm. Then, use the appropriate molar extinction coefficient ($\varepsilon$) to determine the molarity of each ($\varepsilon$ of Cy3 = 150,000 M$^{-1}$ cm$^{-1}$; $\varepsilon$ of Cy5 = 250,000 M$^{-1}$ cm$^{-1}$). To determine the amount of conjugated dye, the labeled samples are diluted 5:1 in Ab microarray desalting buffer. Optical density values for Cy3-labeled samples are read at 552 nm and for Cy5-labeled samples are read at 650 nm. Dye to protein ratios are determined using the following equations:

      [Cy5 Dye] = (A650)/250,000  
      [protein extract] = [BCA value − (0.05 • (A650))]/60000*  
      (Dye/Protein) = [dye]/[protein extract]  
      (Dye/Protein) = (0.24 • (A650))/[BCA value − (0.05 • (A650))]—for Cy5-labeled samples  
      [Cy3 Dye] = (A552)/150,000  
      [protein extract] = [BCA value − (0.08 • (A552))]/60000*  
      (Dye/Protein) = [dye]/[protein extract]  
      (Dye/Protein) = (0.40 • (A552))/[BCA value − (0.08 • (A552))]—for Cy3-labeled samples  
      *is the assumed average molecular weight of the proteins in total extract (*see* **Note 12**).

11. Proceed immediately with *Antibody Array Incubation* (*see* **Subheading 3.3.**)

### *3.2. Small Scale Protocol*

#### *3.2.1. Extracting Protein from Crude Tissue*

1. Before starting, chill the following items on ice or at 4°C:

   - Extraction/labeling buffer
   - one mortar & pestle
   - two 2-ml microcentrifuge tubes
   - one 15-ml conical centrifuge tube

2. Transfer 15–25 mg of frozen tissue to a pre-chilled mortar (*see* **Note 2**).
3. Add 2–5 mg of alumina to the mortar.
4. Use the pestle to grind the tissue until a paste is formed.
5. Add 100–200 µl of pre-chilled extraction/labeling buffer.
6. Mix the buffer into the paste using the pestle. When you finish, use a micropipette tip to scrape the paste that adheres to the pestle back into the mortar.
7. Transfer the extract to a pre-chilled 1.5-ml microcentrifuge tube.
8. While holding the pestle over the mortar, rinse the pestle with 100–200 µl of extraction/labeling buffer.
9. Combine the rinse with the original extract in a 1.5-ml tube.
10. Centrifuge the suspension at 10,000 × *g* for 30 min at 4°C.
11. While taking care not to disturb the pellet, transfer the supernatant to a pre-chilled 1.5-ml microcentrifuge tube.
12. Gently invert the tube to mix the lysate.
13. Measure protein concentration using Pierce's BCA Protein Assay Reagent Kit (*see* **Note 3**).
14. Dilute each sample to 1.1 mg protein/ml by adding the appropriate volume of extraction/labeling buffer. The final volume must be ≥200 µl in order to proceed with labeling of the proteins (*see* **Note 4**).
15. Proceed immediately with the next section.

#### *3.2.2. Labeling Protein with Fluorescent Dye*

Important: Complete **steps 1–10** rapidly without interruption. Once the Cy3 and Cy5 dyes are dissolved in buffer, they must be used immediately.

1. Set up and label one 1.5-ml microcentrifuge tube for each sample: A-Cy3, A-Cy5, B-Cy3, and B-Cy5.
2. Dissolve the Cy3 dye in 110 µl of extraction/labeling buffer by adding the buffer directly to the tube in which the dye is supplied. Note: Each tube of dye contains a quantity of dye that will label approximately 1 mg of total protein.
3. Mix thoroughly by vortexing for 20 s.
4. Centrifuge the tube at moderate speed for 10 s to recover the liquid in the bottom of the tube.
5. Prepare a solution of Cy5 dye in the same manner by following **steps 2–4.**
6. Add 10 µl of Cy3 solution to tubes "A-Cy3" and "B-Cy3."

7. Add 10 µl of Cy5 solution to tubes "A-Cy5" and "B-Cy5."
8. Add 90 µl of protein sample A to tubes "A-Cy3" and "A-Cy5."
9. Add 90 µl of protein sample B to tubes "B-Cy3" and "B-Cy5."
10. Carefully pipette up and down several times to mix the contents. Then, centrifuge each tube at moderate speed for 10 s to recover the liquid in the bottom of the tube.
11. Incubate all four tubes on ice (or at 4°C) for 90 min. Mix by gently vortexing every 20 min.
12. Add 4 µl of blocking buffer to each tube.
13. Incubate each tube on ice (or at 4°C) for 30 min. Mix by gently vortexing every 10 min.

### 3.2.3. Removing Unbound Dye (Desalting) and Determination of Incorporated Dye

The following protocol is intended for use with PD-10 desalting columns, manufactured by Amersham Biosciences. As long as you work quickly, Desalting can be completed at room temperature. Otherwise, if you have access to a cold room, we suggest you complete the procedure at 4°C.

1. Set up and label one PD-10 desalting column and one 2-ml microcentrifuge tube for each sample: A-Cy3, A-Cy5, B-Cy3, and B-Cy5.
2. Prepare 20 ml of 1× desalting buffer for each sample by diluting 10× desalting buffer with the appropriate volume of Milli-Q-grade H2O. Store 1× desalting buffer in a clean plastic tube. Note: Be sure that 1× desalting buffer has a pH = 7.4. Adjust the pH if necessary using dilute HCl or NaOH.
3. Equilibrate each column with 3 × 5 ml of 1× desalting buffer.
4. Apply the Cy3- and Cy5-labeled protein samples (~100 µl each) to the corresponding columns. Allow the protein sample to pass into the column.
5. Add 2.5 ml of 1× desalting buffer to each column. Allow the buffer to pass into the column to push the protein sample further along.
6. Place the 2-ml microcentrifuge tubes (from **step 1**) under the corresponding columns.
7. Elute each protein sample by applying 1.5 ml of 1× desalting buffer to each column. Collect the flowthrough.
8. Store the tubes on ice.
9. Measure protein concentration using the "Test Tube Procedure" in Pierce's BCA Protein Assay Reagent Kit (*see* **Notes 5–9**).
10. Estimate the average number of dye molecules covalently coupled to each protein. Follow the protocol in the Amersham Product Specification Sheet, but first see the instructions below for estimating the average number of coupled dye molecules (*see* **Notes 10** and **11**):

    • Measure Cy3 absorbance at 552 nm and Cy5 absorbance at 650 nm. Then, use the appropriate molar extinction coefficient ($\varepsilon$) to determine the molarity of each

($\varepsilon$ of Cy3 = 150,000 M$^{-1}$ cm$^{-1}$; $\varepsilon$ of Cy5 =250,000 M$^{-1}$ cm$^{-1}$). Optical density values for Cy3-labeled samples are read at 552 nm and for Cy5-labeled samples are read at 650 nm. Dye to protein ratios are determined using the following equations:

[Cy5 Dye] = (A650)/250,000
[protein extract] = [BCA value – (0.05 • (A650))]/60000*
(Dye/Protein) = [dye]/[protein extract]
(Dye/Protein) = (0.24 • (A650))/[BCA value – (0.05 • (A650))]—for Cy5-labeled samples
[Cy3 Dye] = (A552)/150,000
[protein extract] = [BCA value – (0.08 • (A552))]/60000*
(Dye/Protein) = [dye]/[protein extract]
(Dye/Protein) = (0.40 • (A552))/[BCA value – (0.08 • (A552))]—for Cy3-labeled samples
*is the assumed average molecular weight of the proteins in total extract (*see* **Note 12**).

11. Proceed immediately with *Antibody Array Incubation* (*see* **Subheading 3.3.**).

### *3.3. Antibody Array Incubation*

1. Prepare 45 ml of incubation buffer by mixing 4.5 ml of background reducer with 40.5 ml of stock incubation buffer. We provide you with 80 ml of stock incubation buffer. Use 40.5 ml at this step to make "incubation buffer." Store incubation buffer in a clean plastic bottle or tube.
2. Set up the incubation tray provided. Note that it contains four separate chambers for incubating and washing microarrays 1 and 2. You may find it helpful to mark the exterior surface of the tray with a pen to remind you of these assignments: Slide 1 Incubation, Slide 1 Wash, Slide 2 Incubation, Slide 2 Wash.
3. Add 5 ml of incubation buffer to the incubation chambers.
4. Set up two 1.5-ml microcentrifuge tubes. Label the tubes Slide 1 Mix and Slide 2 Mix.
5. In the 1.5-ml microcentrifuge tubes, combine protein samples A and B as follows:

   - Slide 1 Mix: combine 100 µg of protein sample A-Cy5 with 100 µg of protein sample B-Cy3.
   - Slide 2 Mix: combine 100 µg of protein sample A-Cy3 with 100 µg of protein sample B-Cy5 (*see* **Note 13**).

6. Transfer 10–20 µg of protein from the Slide 1 Mix to the Slide 1 incubation chamber. Transfer an equal quantity of protein from the Slide 2 Mix to the Slide 2 incubation chamber (*see* **Note 14**).
7. Incubate the tray at room temperature for 30 min with gentle rocking.

8. Meanwhile, prepare the Ab microarrays by washing the slides two times as follows:
   Caution: Use gloved hands or tweezers to hold and manipulate the microarrays. Never touch the array-end of the slide. Instead, always hold the slide at the end nearest the affixed label.

   a. While pressing your gloved finger against the top of the vial to keep the slides from falling out, decant the storage buffer from the green-capped storage vial. Note: The storage buffer contains glycerol and should be disposed of in a properly labeled waste container.
   b. Add 30 ml of stock incubation buffer.
   c. Cap the storage vial. Then slowly invert the vial 10 times.
   d. Decant the stock incubation buffer while using your gloved finger to keep the slides from falling out.
   e. Add 20 ml of incubation buffer (prepared in **step 1**).
   f. Repeat **step c**.
   g. Stand the vial upright in a rack.

9. Record each slide's lot number and assign one slide to the Slide 1 Mix and one slide to the Slide 2 Mix.
10. Remove the slides one by one from the storage vial and place each, arrayside-up, in the tray chamber containing the incubation buffer/slide mix to which it has been assigned. (The array is printed on the side to which the label is affixed.)
11. Incubate the slides at room temperature for 30 min with gentle rocking. Every 10 min, perform the following manipulation to assist the exchange of liquid on all sides of the slide: Use a micropipette tip to pry up one end of the slide while you gently rock the Incubation Tray once or twice.
12. Add 5 ml of incubation buffer (prepared in **step 1**) to each wash chamber.
13. Transfer the slides to their respective wash chambers (*see* **Note 15**).
14. Incubate at room temperature for 5 min with gentle rocking (*see* **Note 16**).
15. Remove the buffer from the wash chambers.
16. Add 5 ml of wash buffer 1.
17. Incubate at room temperature for 5 min with gentle rocking.
18. Repeat **steps 15–17** using wash buffer 2; then using wash buffer 3. And so on, until you have washed each slide with each of the wash buffers 1–7.
19. Dry the slides. It is important to remove as much moisture as possible from the surface of the slides before the water evaporates passively. We recommend the following method:

    a. Using gloved hands and holding the slides by their edges only, place the slides, array-end-up, in the empty, green-capped storage vial provided. Important: Do not touch the array surface.

   b. Cap the vial and centrifuge the slides at $1000 \times g$ for 25 min at room temperature (*see* **Note 17**).

   c. Using gloved hands, uncap the vial. While holding your finger over the top of the vial to prevent the slides from falling out, tip the vial slightly to nudge the slides near the rim of the vial. When the slides protrude by approximately 2 cm, remove the slides one by one.

Important: Do not touch the array surface when removing the slides. Instead, hold the slides by their edges.

20. Scan the slides with a microarray scanner. If you need to postpone the scanning, keep the slides in a dry chamber and protect them from light until you are ready to scan.

21. Scan slides $\geq 24$ h after drying (*see* **Notes 18–20**).


### *3.4. Analysis of Results*

To use our Ab Microarray Analysis Workbook as described below, you must first calculate the Cy5/Cy3 fluorescent signal ratios for all coordinates on each array. This calculation can usually be done with your array analysis software (e.g., GenePix Pro). The Cy5/Cy3 values are required to calculate INRs, as described below. The Ab Microarray Analysis Workbook is a Microsoft® Excel 97/98 file that converts your fluorescence data into INRs for each coordinate on the array. As described in the *Introduction*, the INR calculated by our workbook is a numerical value that represents the abundance of antigen in sample A relative to that of sample B.

1. Connect to http://bioinfo.clontech.com and download a copy of the workbook that corresponds to the Slide Lot Number of your microarray. The Microarray Slide Lot Number is given on the data label affixed to the glass slide (*see* **Note 21**).

2. Launch Microsoft Excel. Then, open the Microarray Analysis Workbook. On opening the workbook, you will notice that it contains four worksheets. The names of these sheets appear on tabs at the bottom of the workbook window.
The "Ab," "Array," and "Ab List" worksheets contain array-specific information such as the names and coordinates of antibodies and the Locus Link and SWISS-PROT accession numbers of the corresponding protein targets.

3. The fourth worksheet, "Import & Analyses," contains formulas that perform arithmetic operations on the fluorescence data (i.e., Cy5/Cy3 signal ratios) that you paste into the worksheet. Other formulas in this sheet combine the values of these operations to generate an INR for each coordinate on the array (see **Note 22**).
An example of generated data is shown below comparing human heart and liver tissue extracts (*see* **Fig. 4**).
Some of the western blot data showing both confirmed and disagreeing results is presented in **Fig. 5** (*see* **Note 23**).

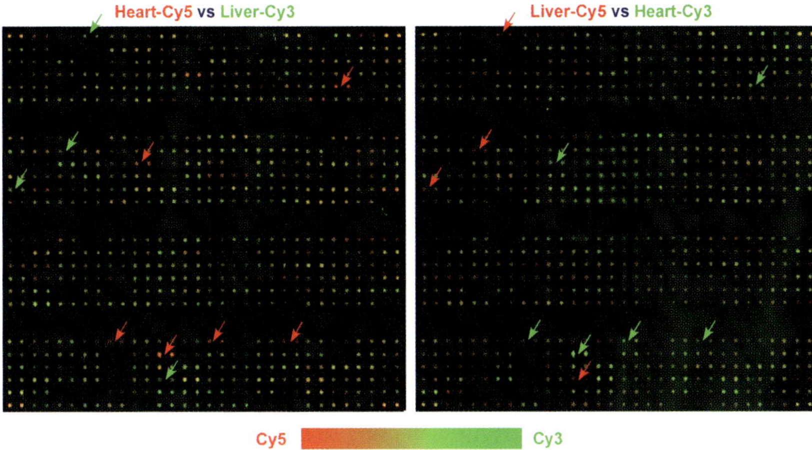


Fig. 4. Pseudo-color presentation of signals generated on the antibody microarray using dual color/reverse color experimental setup. Ten microgram of labeled protein per channel was incubated with two arrays using the dual color/reverse color setup. Red arrow on the first slide and green arrows on the second slide point out to some antigens with higher abundance in the heart sample, whereas green arrows on the first slide and red arrows on the second slide point out to some antigens with higher abundance in the liver sample (*see* **Table 2**).

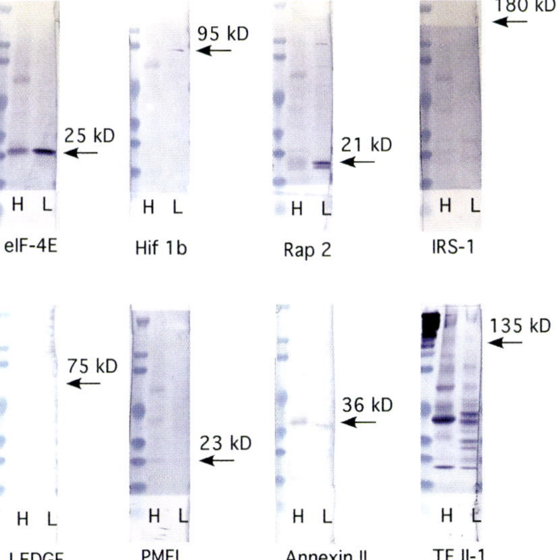


Fig. 5. Partial western validation results of the data obtained during comparison of human heart and liver tissues (*see* **Fig. 4** and **Table 2**). The detected differences were validated by western, and we found more than 80% correlation with the array findings.

**Table 2**
**Relative Abundance Differences of Proteins in Liver and Heart**

| Higher in liver | | Higher in heart | |
|---|---|---|---|
| Antigen name | INR | Antigen name | INR |
| Acetylcholinesterase | 2.8 | HPV-16 L1 | 10.9 |
| Topoisomerase II a | 2.5 | LEDGFNTF2 | 3.9 |
| DNA Polymerase d | 2.2 | Ku70 | 3.3 |
| HIF-1b | 2.1 | Inhibitor 2 | 2.8 |
| Rap 2 | 2.0 | CDC27 | 2.7 |
| Pericentrin | 2.0 | Annexin XI | 2.4 |
| cdk1 | 2.0 | Caspase7/Mch3 | 2.3 |
| eIF-4E | 1.9 | p57/Kip2 | 2.3 |
| GAGE | 1.9 | DEK | 2.2 |
| GFAP | 1.8 | Hsp 70 | 2.2 |
| PECI | 1.8 | IKKa/1 | 2.1 |
| IRS-1 | 1.7 | EGF Receptor | 2.0 |
| Transportin | 1.7 | MKK7 | 1.9 |
| TLS | 1.7 | Adaptin d | 1.9 |
| Cdk7 | 1.7 | DDX1 | 1.8 |
| VHL | 1.7 | Integrin b3 (CD61) | 1.8 |
| MLH1 | 1.7 | Ubc9 | 1.8 |
| Dynamin II | 1.6 | p96 | 1.8 |
| K+ Channel a | 1.6 | TFII-I/BAP-135 | 1.7 |
| Btf | 1.6 | XPA | 1.7 |
| Rb2 | 1.6 | Neurexin I | 1.6 |
| ZFP-37 | 1.6 | PMF-1 | 1.6 |
| MGMT | 1.6 | Gap1m | 1.6 |
| Tau-5 | 1.6 | Bog | 1.6 |
| XIN | 1.6 | GOK/Stim1 | 1.6 |
| Ezrin | 1.6 | Ankyrin B | 1.6 |
| Retinol-binding protein | 1.5 | TRF2 | 1.5 |
| PKC q | 1.5 | Apo E | 1.5 |
| Endoglin | 1.5 | Rin1 | 1.5 |
| Cathepsin L | 1.5 | Bcl-x | 1.5 |
| MyoD | 1.5 | iNOS/Type II | 1.5 |
| Nur77 | 1.5 | Stat 2 | 1.5 |
| NFAT-1 | 1.5 | EBP50 | 1.5 |

## 4. Notes

1. For first time users, we always recommend using the large scale protocol first: One will end up with a lot of leftover extracted protein and labeled protein, but there is a better chance of getting a full representation of proteins in your sample.

2. Weigh the tissue sample in the mortar (pre-weigh the mortar to determine the exact amount of the tissue sample) *before* adding the buffer; Extraction/labeling buffer is added at volume/weight ratio of 20 µl/mg of tissue.

3. Use the correct Pierce BCA kit (Pierce Biotechnology; Cat. No. 23225 or 23227) for determination of the protein concentration; the other kits will be affected by the components of our buffers. Pierce's BCA Protein Assay Reagent Kit should be used for all Ab microarray analyses. Using other BCA reagents (or kits) could lead to errors in protein estimation, because Ab microarray buffers contain substances known to interfere with protein assays. Pierce's Kit has been tested by our scientists and approved for use with Ab microarray procedures and reagents.

4. At least 5% yield of protein has to be achieved from the starting tissue weight, otherwise extraction is incomplete—most likely there is a biased extraction of high-abundance proteins causing both lower sensitivity and misrepresentation of protein ratios. Repeat the extraction procedure and make sure that the grinding results in complete homogenization of the sample.

5. Do not dilute the samples further for the steps below as the concentration of the protein in the samples is already low because of the dilution resulting from the desalting step.

6. We recommend you use the BCA Protein Assay Reagent Kit from Pierce Biotechnology. It has been tested by our scientists and shown to be compatible with our buffers. Using other BCA reagents (or kits) could lead to errors in protein estimation.

7. 'Use BSA as your protein standard. Construct a standard curve using BSA solutions that are 0.02, 0.05, 0.1, 0.2, 0.3, 0.4, and 0.5 mg/ml.

8. We recommend you measure each standard and protein sample in triplicate. Use 1× desalting buffer as a blank (i.e., as the 0 mg/ml sample).

9. Because Cy3 and Cy5 absorb at 562 nm, you will need to subtract the dyes' contribution to the overall OD562. To do this, prepare a protein blank that contains an aliquot of your labeled protein sample and 1× desalting buffer, substituted for BCA reagent. Calculate the $\Delta OD562 = (OD562$ Protein sample $- OD562$ Protein Blank). Use $\Delta OD562$ to calculate the protein concentration utilizing the standard curve obtained from the BSA standards.

10. Unless you have more precise measurements for the distribution of protein mass in your sample, assume that the average molecular weight of protein is 60 kDa.

11. Use the value determined with Pierce's BCA Protein Assay Reagent Kit, not the A280 method, to calculate the molar concentration of protein in your sample.

12. Best results are obtained when the dye : protein ratio is between 2 and 4 for all four labeling reactions. If the dye : protein ratio is above 6, repeat labeling by titrating out the amount of dye that over-labeled:
Example: standard label uses 1:9 dye : protein volume (50 µl dye: 450 µl protein at protein concentration of 1.1 mg/ml). Use lower ratio of dye to protein:

    40 µl dye +10 µl desalting buffer + 450 µl protein
    35 µl dye+ 15 µl desalting +450 µl protein

    There should be less than 50% difference in the incorporated Cy3 in both samples and Cy5 in both samples. There should be less than two times difference in the incorporation levels between the Cy3-labeled samples and the Cy5-labeled samples. If any of these criteria are not met, one should make sure that the starting amount of the protein is the same in both samples and repeat the labeling reaction after making sure that all the dye in the Cy3 and Cy5 tubes has been homogenously solubilized before transferring aliquots to the protein samples.

13. If desired, the remainders of samples A and B can be stored at 4°C (short-term storage) or –20°C (long-term storage) for later use in other applications—e.g., western blotting. We do not recommend you use stored protein samples for future microarray analyses as the incorporation of dye can influence the solubility of the proteins upon storage.

14. For samples derived from serum, use 50–100 µg of protein, because a large majority (~80%) of the protein consists of immunoglobulin and albumin.

15. Do not let the slides *dry* during the incubation and washing procedure.

16. Do not wash more than 5 min per washing step—binding of the antigen to the antibody is an equilibrium process—as the concentration of the antigen in the washing buffer is very low, extensive washing results in more antigen being released from the antibody–antigen complex into the liquid phase resulting in decreased detected signal.

17. The spin/drying step might result in leftover liquid, which could cause streaks and background on slide resulting in lower signal to noise ratio. A couple of ways to minimize this are to poke holes in the bottom of the empty storage chamber to allow the liquid to escape *or* use Falcon 50 ml conical with Kimwipes at bottom to absorb the liquid.

18. We routinely scan the arrays with an Axon GenePix 4000B scanner using the following settings:

    - 635 nm (Cy5 channel): PMT = 670 V; power = 33%.
    - 532 nm (Cy3 channel): PMT = 550 V; power = 33%.

    The laser power should not be set too high or uneven between the two detection wavelengths. Many people have a hard time setting the laser at 30% gain, as oligo arrays require 100% power. One can use as a rough indication the pre-labeled Cy3/Cy5 orientation markers. They should have a similar value (within 25% variation) and preferably be around 3000–10,000 mean fluorescence units

(FU). In addition, the background FU should be around 100–200 mean FU. The power (voltage) and gain should be at the same levels for data acquisition from both slides. If for some reason a change in the settings is done for the second slide, the signals from the first slide have to be collected again using the new settings.

19. The BSA control spots can be used as a guide for setting the scanner. These control spots should generally give a fluorescent signal of 2500–30,000 FU. If the control spots are >50,000 FU, this is an indication that the scanner settings are too high; the slides should be rescanned using lower power settings. If you are using a scanner other than the GenePix 4000B instrument, adjust the laser power and PMT (if possible) to obtain a signal within the range of 2500–30,000 FU for the control spots.

20. Some scanners (such as the PerkinElmer ScanArray instrument) adjust the laser settings based on a pre-scan of the slide; these types of scanners generally calculate the correct settings so adjustments are not normally required. However, it is still useful to check the signal intensities of the control spots as the automatic detection settings based on the pre-scan may be incorrect.

21. We supply Gal files that work both for Axon and Perkin-Elmer scanners. Export the data as tab-delimited text file and import the columns with the mean fluorescent signal in the worksheet we supply.

22. In the last step, calculate the mean INR for all of the 507 antibodies that have significant signal to noise ratio (S/N $\geq 2$). Identify the targets with significant abundance values that are with INR:
    Lower than mean INR $\times$ 0.7
    Higher than mean INR $\times$ 1.3

23. It is clear from this data that sometimes it is difficult to interpret the data from the Western—for an example, Insulin Receptor Substrate 1 (IRS-1) was detected in higher abundance in liver; however, the Western did not reveal any protein bands at 180 kDa, and on overall, one could consider the total signal generated in the lane that was loaded with the heart extract to be higher than in the liver band (denoted with H for heart and L for liver). The opposite is true for the TF II-1/Bap 135 Western results, where the array data showed increased amount of the antigen in the heart sample, but we detected a very strong band around 50 kDa instead of 135 kDa. There were few Westerns that showed data that was directly contradictory to the array results (not shown). This is to be expected when two very different methods for preparation of the sample and detection (one with native protein and the other with denatured protein) are used. For this reason, we believe that all targets identified with the array have to be validated by alternative means.

## Acknowledgments

I express my gratitude to Inger Larsen and Joshua Ehrlich for their help with the linguistic and formatting review.

## References


1. Aebersold, R., Rist, B. and Gygi, S.P. Quantitative proteome analysis: Methods and applications. *Ann N Y Acad Sci* 919, 33–47 (2000).
2. Chen, G., et al. Discordant protein and mRNA expression in lung adenocarcinomas. *Mol Cell Proteomics* 1, 304–313 (2002).
3. Haab, B.B. Advances in protein microarray technology for protein expression and interaction profiling. *Curr Opin Drug Discov Devel* 4, 116–123 (2001).
4. Arenkov, P., et al. Protein microchips: Use for immunoassay and enzymatic reactions. *Anal Biochem* 278, 123–131 (2000).
5. Bussow, K., Konthur, Z., Lueking, A., Lehrach, H. and Walter, G. Protein array technology. Potential use in medical diagnostics. *Am J Pharmacogenomics* 1, 37–43 (2001).
6. Huang, R.P., Huang, R., Fan, Y., and Lin, Y. Simultaneous detection of multiple cytokines from conditioned media and patient's sera by an antibody-based protein array system. *Anal Biochem* 294, 55–62 (2001).
7. Lueking, A., Horn, M., Eickhoff, H., Bussow, K., Lehrach, H., and Walter, G. Protein microarrays for gene expression and antibody screening. *Anal Biochem* 270, 103–111 (1999).
8. Robinson, W.H., Steinman, L., and Utz, P.J. Protein and peptide array analysis of autoimmune disease. *Biotechniques* (Suppl), Dec 66–69 (2002).
9. de Jager, W., te Velthuis, H., Prakken, B.J., Kuis, W., and Rijkers, G.T. Simultaneous detection of 15 human cytokines in a single sample of stimulated peripheral blood mononuclear cells. *Clin Diagn Lab Immunol* 10, 133–139 (2003).
10. Bock, C., et al. Photoaptamer arrays applied to multiplexed proteomic analysis. *Proteomics* 4, 609–618 (2004).
11. Shao, W., et al. Optimization of rolling-circle amplified protein microarrays for multiplexed protein profiling. *J Biomed Biotechnol* 2003, 299–307 (2003).
12. Anderson, K., Potter, A., Baban, D., and Davies, K.E. Protein expression changes in spinal muscular atrophy revealed with a novel antibody array technology. *Brain* 126, 2052–2064 (2003).
13. Marienfeld, C., et al. Translational regulation of XIAP expression and cell survival during hypoxia in human cholangiocarcinoma. *Gastroenterology* 127, 1787–1797 (2004).
14. Yamagiwa, Y., Marienfeld, C., Meng, F., Holcik, M., and Patel, T. Translational regulation of x-linked inhibitor of apoptosis protein by interleukin-6: A novel mechanism of tumor cell survival. *Cancer Res* 64, 1293–1298 (2004).
15. Hudelist, G., Pacher-Zavisin, M., Singer, C.F., Holper, T., Kubista, E., Schreiber, M., Manavi, M., Bilban, M., and Czerwenka, K. Use of high-throughput protein array for profiling of differentially expressed proteins in normal and malignant breast tissue. *Breast Cancer Res Treat* 86, 281–291 (2004).
16. Ghobrial, I.M., McCormick, D.J., Kaufmann, S.H., Leontovich, A.A., Loegering, D.A., Dai, N.T., Krajnik, K.L., Stenson, M.J., Melhem, M.F., Novak, A.J.,

Ansell, S.M., and Witzig T.E. Proteomic analysis of mantle cell lymphoma by protein microarray. *Blood* 105, 3722–3730 (2005).

17. Gosmanov, A.R., Umpierrez, G.E., Karabell, A.H., Cuervo, R., and Thomason, D.B. Impaired expression and insulin-stimulated phosphorylation of Akt-2 in muscle of obese patients with atypical diabetes. *Am J Physiol Endocrinol Metab* 287, E8–E15 (2004).

# 10

# SH2 Domain-Based Tyrosine Phosphorylation Array

**Xin Jiang, Lesile Roth, Stephanie Han, and Xianqiang Li**

## Summary

Tyrosine phosphorylation plays a major role in intracellular signal transduction pathways. Phosphorylated tyrosine residues initiate signaling pathways by recruiting proteins containing Src homology-2 (SH2) domains. Herein is described a high-throughput assay to detect interactions between phosphorylated proteins and human SH2 domains, in order to profile the genome-wide phosphorylation status of cells under a variety of biological conditions. The SH2 domain array was prepared by immobilizing 115 SH2 domain proteins on a membrane. The array assay is straightforward with no expensive equipment or radioactivity required: the cell lysate is incubated with the array membrane, and the signal is measured using a chemiluminescence-based detection system.

**Key Words:** SH2 domain; tyrosine phosphorylation; array; chemiluminescence; high throughput.

## 1. Introduction

Protein phosphorylation is a post-translational modification at tyrosine or serine/threonine residues that regulate a number of critical biological processes. Tyrosine phosphorylation provides control of complex physiological events, including extracellular signal recognition, ion fluxes, cell motility, gene expression, cell proliferation, and programmed cell death. It is estimated that only a tiny fraction, about 0.1%, of total protein phosphorylation is mediated through tyrosine *(1,2)*; however, a slight change in protein tyrosine phosphorylation signaling can cause major alterations in many regulatory processes and ultimately result in cancer or other human diseases.

Tyrosine phosphorylation is mediated by protein tyrosine kinases (PTKs), and more than 90 PTK genes have been identified in the human genome *(3)*.

From: *Methods in Molecular Biology, vol. 441: Tissue Proteomics: Pathways, Biomarkers, and Drug Discovery*
Edited by: B.C.-S. Liu © Humana Press, Totowa, NJ

One of the most important groups of PTKs is the receptor tyrosine kinases (RTKs). Among the 90 PTKs identified, 58 are RTKs *(4)*. RTKs are activated by the binding of specific ligands, which leads to the dimerization of the receptor and autophosphorylation of specific receptor tyrosine residues.

In general, tyrosine autophosphorylation either stimulates the intrinsic catalytic (kinase) activity of the receptor or generates recruitment sites for downstream signaling proteins containing phosphotyrosine-recognition domains such as Src homology-2 (SH2) and protein tyrosine binding (PTB) domains. SH2 domains are protein-binding domains that mediate signal transduction pathways associated with phosphotyrosine kinase by interacting with phosphotyrosine residues of target proteins *(5)* [reviewed in *(6–8)*]. It has been estimated that there are 115 SH2 domains encoded by the human genome *(9)*. SH2 domains, found in proteins with diverse biochemical functions, including growth factor receptors and adaptor proteins, play a critical role in phosphotyrosine-dependent signaling by connecting receptors and downstream signaling molecules to adaptor molecules. This linkage allows signals, such as those that regulate kinase activity, to be transmitted within a cell. Mutations within the SH2 domains of a number of proteins have been causally implicated in human disease.

A number of techniques have been applied to profile tyrosine phosphorylation status *(10)* [reviewed in *(4)*]. Most existing approaches are based on a mass-spectrometry (MS) detection system, which is low-throughput, time-consuming, and labor intensive [reviewed in *(4)*]. Recently, the antibody array has been used for profiling the expression of tyrosine phosphorylated proteins. However, it does not provide information on other important aspects of tyrosine phosphorylation, such as pathway interaction partners or associated signaling pathways.

Here, we introduce a high-throughput assay for monitoring genome-wide tyrosine phosphorylation by detecting the interaction of all 115 human SH2 domains, immobilized on a membrane, with tyrosine-phosphorylated proteins present in cell lysate samples (*see* **Fig. 1**). The array assay can profile and monitor tyrosine phosphorylation in cells under a variety of biological conditions.

## 2. Materials

1. Glutathione-S-transferase (GST) expression vector. pGST-p (Panomics, Fremont, CA, USA) or pGEX-4T (GE Healthcare, Piscataway, NJ, USA).
2. Glutathione Superflow Resin and purification system (Clontech Laboratories, Mountain View, CA, USA).
3. Full-length cDNA amplification (Panomics).

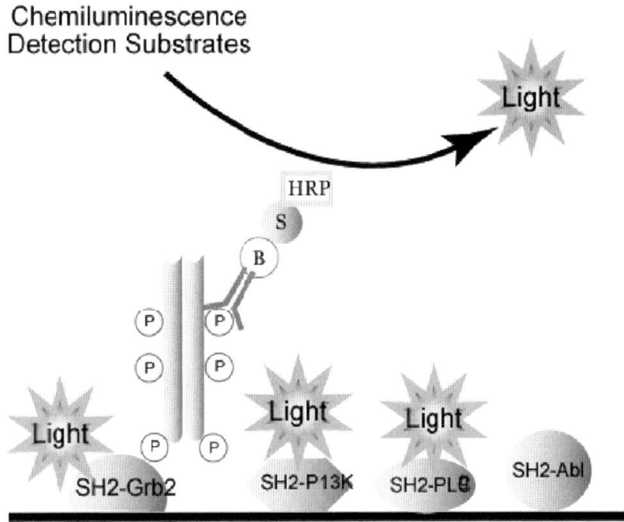


Fig. 1. Diagram illustrates the principle of Src homology-2 (SH2) domain array for genome-profiling tyrosine phosphorylation. There are three basic steps: 1. SH2 domains are spotted on Polyvinylidene fluoride (PVDF) membrane; 2. Cell lysate is incubated with array; 3. The phosphorylated proteins are detected with phosphotyrosine antibody.

4. Restriction enzymes (New England Biolabs).
5. T4 DNA ligase (Roche Applied Science, Indianapolis, IN, USA).
6. *Escherichia coli* strains DH5α.
7. LB: 1% tryptone, 0.5% yeast extract, 1.0% sodium chloride, and/or 1.5% agar).
8. 2YT broth:1.6% Tryptone, 1.0% yeast extract, and 0.5% sodium chloride.
9. Ampicillin.
10. Isopropyl β–D-thiogalactopyranoside (IPTG).
11. Incubator.
12. Sonicator.
13. Tissue culture hood.
14. $CO_2$ incubator.
15. PMSF (Sigma Aldrich, St. Louis, MO, USA).
16. Cell culture medium [Dulbecco's modified Eagle's medium (DMEM) plus 10% serum].
17. Orbital shaker.
18. Blocking buffer (SuperBlock, Pierce, Rockford, IL, USA).
19. Resuspension Buffer (1× PBS).
20. Wash Buffer (24.7 mM Tris, pH 7.4, 2.7 mM potassium chloride, 0.137 M sodium chloride, 0.05% Tween-20).
21. Polyvinylidene fluoride (PVDF) Membrane, such as Immobiolon-P (Millipore Corp., Bedford, MA, USA).
22. TranSignal™ SH2 Domain Array (Panomics, Inc.).

23. Anti-histidine HRP Conjugate, such as Anti-Polyhistidine Horse Radish Peroxidase (HRP) Conjugate (Sigma-Aldrich; St. Louis, MO, USA).
24. 2× Cell Lysis Buffer: 40 mM Tris pH 7.5, 300 mM NaCl, 2 mM ethylenedi-amine-tetraacetic acid (EDTA), 2% Triton X-100, 5 mM sodium pyrophosphate, 2 mM glycerophosphate, 2 mM Na3VO4, 2 µg/ml leupeptin.
25. Protease inhibitor cocktail (Sigma, Cat. No. P2850).
26. PMSF (Sigma, Cat. No. P7626).
27. Phosphotyrosine RC20 monoclonal antibody Biotin conjugate (Pharmingen, Cat. No. 610021).
28. Detection reagents—peroxide solution and luminol enhancer—such as ECL reagents (Amersham Biosciences) or SuperSignal West Pico Chemiluminescent substrates (Pierce).
29. Film and developer—such as Hyperfilm ECL (Amersham Biosciences)—or a chemiluminescence imaging system—such as the FluorChem imager (Alpha Innotech Corp., San Leandro, CA, USA).

## 3. Methods

The methods described below outline the preparation of recombinant SH2 domain proteins in *E. coli* (*see* **Subheading 3.1.**), the preparation of an SH2 domain array (*see* **Subheading 3.2.**), the preparation of cell lysate (*see* **Subheading 3.3.**), incubation of the cell extracts with the SH2 domain array (*see* **Subheading 3.4.**), and detection (*see* **Subheading 3.5.**). If a prepared array is purchased, start as given in the steps of **Subheading 3.3.**

### *3.1. Preparation of the Recombinant SH2 Domain Proteins in E. coli*

The preparation of the recombinant GST-tagged SH2 domain proteins follows standard molecular biology-cloning techniques *(8)*. A brief summary follows:

1. Design and synthesize specific SH2 primers and amplify SH2 cDNA from full-length cDNA pool. The names of SH2-containing proteins are listed in **Table 1**.
2. Insert human SH2 domain cDNAs into the multiple cloning sites of GST expression vector (Amersham Biosciences). This will express GST fusion proteins that are transcriptionally initiated from the P*lac*-IPTG inducible promoter.
3. Transform the recombinant clones into *E. coli* DH5α
4. Prepare a bacterial culture expressing recombinant proteins by diluting an overnight culture 1:100 with 2YT medium (plus ampicillin at 100 µg/ml), growing at 30 °C until OD600 reached 0.5–1, and then adding IPTG to a final concentration of 0.1–1.0 mM and continuing incubation for 4 h at 37ºC.
5. Collect cells by centrifugation, disrupt by sonication, then purify the recombinant proteins according to the supplier's recommendation (*see* **Note 1**).

**Table 1**
**The List of Genome-Wide Sre homology-2 (SH2) Domain-Containing Proteins**

|   | 1 | 2 | 3 | 4 | 5 | 6 | 7 | 8 | 9 | 10 |
|---|---|---|---|---|---|---|---|---|---|----|
| A | SH3BP2 | ABL1 | ABL2 | HSH2D | APS | BCAR3 | BLK | BLNK | BMX | Pos. |
| B | BRDG1 | BTK | CBL | CBLB | CBLC | CHN1 | CHN2 | CISH | CRK | Pos. |
| C | CRKL | CSK | TNS4 | DAPP1 | SH2D1B | FER | FES | FGR | FRK | Pos. |
| D | FYN | GRAP | GRB10 | GRB14 | GRB2 | GRB7 | GRAP2 | HCK | ITK | Pos. |
| E | JAK1 | JAK2 | JAK3 | LCK | LCP2 | LNK | LYN | MATK | MIST | Pos. |
| F | NCK1 | NCK2 | SH2D3A | PIK3R3-D1 | PIK3R3-D2 | PIK3R1-D1 | PIK3R1-D2 | PIK3R2-D1 | PIK3R2-D2 | Pos. |
| G | PLCG1-D1 | PLCG1-D2 | PLCG2-D1 | PLCG2-D2 | PTK6 | PTPN11-D1 | PTPN11-D2 | PTPN6-D1 | PTPN6-D2 | Pos. |
| H | RaLP | RIN1 | RIN2 | RIN3 | RSG1-D1 | RSG1-D2 | SH1A | SH2A | SH2B | Pos. |
| I | SH2D3C | SHB | SHC1 | SHC2 | SHC3 | SHD | SHF | SHIP | SHIP2 | Pos. |
| J | SLA1 | SLA2 | SOCS1 | SOCS2 | SOCS3 | SOCS4 | SOCS5 | SOCS6 | SOCS7 | Pos. |
| K | SRC | SRMS | STAP2 | STAT1 | STAT2 | STAT3 | STAT4 | STAT5A | STAT5B | Pos. |
| L | STAT6 | SUPT6H | SYK-D1 | SYK-D2 | TEC | TENC1 | TENS1 | TNS1 | TXK | Pos. |
| M | TYK2 | VAV1 | VAV2 | VAV3 | YES | ZAP70-D1 | ZAP70-D2 |  | Pos. |  |
| N | Pos. | Pos. | Pos. | Pos. | Pos. | Pos. | Pos. | Pos. | Pos. | Pos. |

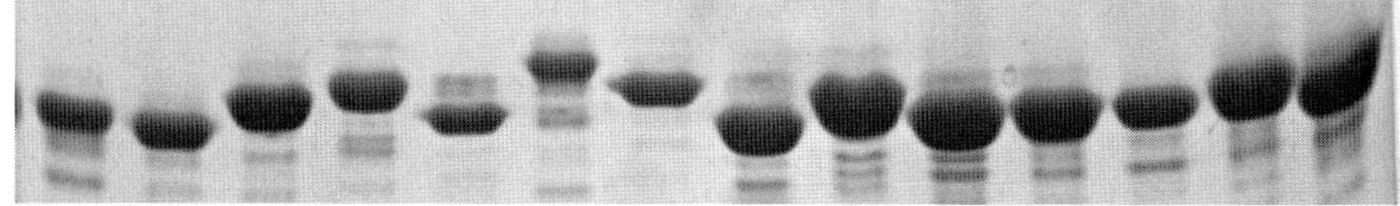

Fig. 2. The representative results for checking quality of recombinant proteins checked by sodium dodecyl sulfate–polyacrylamide gel electrophoresis (SDS–PAGE) gel. Two microgram of SH2 recombinant proteins were loaded on 10% SDS–PAGE gel, and the gels were stained with Coomassie Brilliant Blue.

6. The purified proteins should be checked by sodium dodecyl sulfate–polyacrylamide gel electrophoresis (SDS–PAGE) gel (*see* **Fig. 2**)

## *3.2. Preparation of SH2 Domain Array*

The proteins are purified using standard methodology *(10)*.

1. Determine the amount of purified recombinant protein by a colorimetric assay.
2. Dilute the protein to the same concentration and resolve by SDS–PAGE to determine protein quality and confirm quantity.
3. Immobilize the purified proteins at a standard amount (e.g., 400 and 800 ng) onto PVDF membrane (*see* **Notes 2** and **3**), which has been pre-treated according to manufacturer's recommendation.

## *3.3. Preparation of Mammalian Cell Lysate*

A 10-cm plate of cells at 90% confluence is required per membrane. This should yield between 1 and 5 mg of protein. The final concentration of the cell lysate should be 0.5–4 mg/ml.

1. Before cells are treated with or without stimuli of interest, cells are starved with 0.1% serum in appropriate culture medium for overnight.
2. After appropriate cell treatments, wash adherent cells with ice-cold PBS, and drain. Wash non-adherent cells with ice-cold PBS and centrifuge at 1600 × *g* for 5 min to pellet the cells and remove the PBS (~100 µl leftover).
3. Add ice-cold 100 µl 2× Cell Lysis Buffer to the cells (Panomics). Scrape adherent cells off tissue culture dish with a plastic cell scraper; transfer cell suspension to a 1.5-ml microcentrifuge tube. Maintain all components on ice.
4. Sonicate cell suspension for 5–10 s to shear DNA and reduce sample viscosity.
5. Centrifuge at 7826 × *g* for 5 min and transfer the supernatant to a fresh tube as the cell lysate. The lysate can be used immediately or stored at –80 °C for future use.

### 3.4. Incubation of the Cell Lysate with SH2 Domain Array

In this section, the cell lysate (prepared as in **Subheading 3.3.**) will be incubated with the array membrane prepared in **Subheading 3.2** or purchased (Panomics). In each step, use enough buffers to fully submerge the membrane. Never let the membrane dry out.

1. Place each membrane into a tray containing 5 ml of 1× Wash Buffer for 30 min.
2. Discard the Wash Buffer and add 5 ml of 1× Blocking Buffer. *Make sure that the membrane is fully submerged in buffer.*
3. Place the tray on a shaker and incubate at room temperature for approximately 1–2 h, or until each membrane appears uniformly wetted and no dry spots are visible.
4. Dilute the cell lysate samples, prepared as in **Subheading 3.3**, to 3.0 ml with 1× PBS, per membrane used.
5. Remove 1× Blocking Buffer from the membranes and rinse the membranes once in 1× Wash Buffer.
6. Add the diluted cell lysates to the membrane, cover each tray, and incubate with gentle shaking for between 3 h and overnight at 4 °C.
7. After incubation, wash the membrane three times with 5 ml of 1× Wash Buffer for 10 min (each wash) at room temperature.
8. Add 5 µl detection antibody to 2.5 ml 1× Antibody Dilution Buffer, per membrane used.
9. Transfer 2.5 ml of the diluted detection antibody to each tray and incubate for 2 h at room temperature with gentle shaking (*see* **Note 4**).
10. After incubation, wash the membrane three times with 5 ml of 1× WashBuffer for 10 min (each wash) at room temperature. Decant the wash and proceed to the next incubation step.
11. Add 4 µl Streptavidin-HRP to 4 ml 1× Wash Buffer, per membrane used.
12. Transfer 4 ml of the diluted Streptavidin-HRP to each tray and incubate for 1 h at room temperature with gentle shaking.
13. After incubation, wash the membrane three times with 5 ml of 1× Wash Buffer for 10 min (each wash) at room temperature.

### 3.5. Detection

Do not let the membrane dry out during detection.

1. Prepare the detection solution immediately before use by mixing equal amounts of Detection Buffers (peroxide solution and luminol enhancer)—e.g., 1 ml of peroxide solution and 1 ml of luminol enhancer.
2. Using forceps to hold the membrane at the corner, carefully remove membrane from its tray. Drain the excess Wash Buffer from the membrane by touching the edge against tissue. Place protein-side-up by orienting the notch to the top, right-hand corner on a clean plastic sheet.

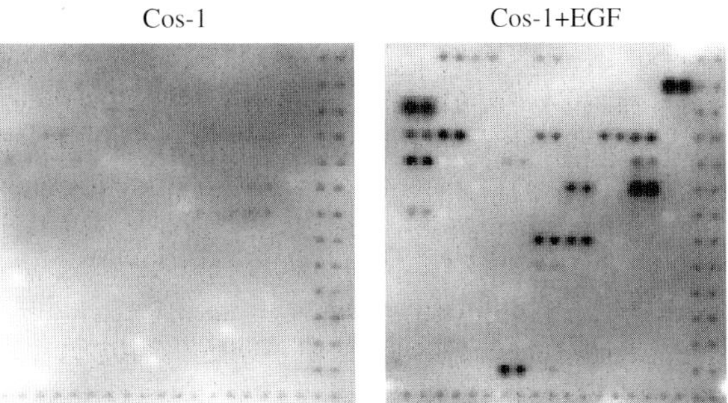



Fig. 3. Genome profiling of epidermal growth factor (EGF)-induced tyrosine phosphorylation in Cos-1. Cos-1 cells were starved with full Dulbecco's modified Eagle's medium (DMEM) with 0.1% serum for overnight. The Cos-1 cells then were treated with or without 20 ng/ml EGF for 30 min. The cells were collected in 1× cell lysis buffer and subjected to array assay. A: Cos-1; B: Cos-1 + EGF treatment.

3. Pipette the mixed Detection Buffers onto the membrane. Ensure that the buffer mixture is evenly distributed over the membrane without air bubbles.
4. Incubate at room temperature, uncovered, for 5 min.
5. Remove excess substrate by holding the membranes with forceps and touching the edge against tissue. Place the membrane between two plastic sheets and gently press on the top sheet to remove air bubbles.
6. Expose the membranes using either Hyperfilm ECL or a chemiluminescence imaging system, such as the FluorChem imager (Alpha Innotech Corp.). In either case, try several different exposures of varying lengths of time (e.g., 30 s–5 min) (*see* **Fig. 3**).

## 4. Notes

1. Some protein arrays are prepared from whole bacterial lysate, but it is not the method of choice for the SH2 domain array. Although the sequence of SH2 domains are conserved in humans, expression levels of each clone in *E. coli* vary. To generate array membranes for a binding assay that results in semi-quantitative or comparative data, it is better to make membranes using purified proteins.
2. Membranes made of PVDF are better suited for this type of protein domain array than nitrocellulose. Nitrocellulose yields higher background signals, which could be interpreted as false positives.
3. Membranes can be stripped and reused, but it is not recommended. Stripping may cause damage to proteins on the array membrane affecting the results of the binding assay.
4. SH2 domain array can be used for profiling general tyrosine phosphorylation with the antibody against phosphotyrosine and detecting specific tyrosine kinase phosphorylation with the antibody against specific active kinase.

# References


1. Blume-Jensen, P., and Hunter, T. (2001) Oncogenic kinase signalling. *Nature* **411**, 355–365.
2. Cooper, J. A., Sefton, B. M., and Hunter, T. (1983) Detection and quantification of phosphotyrosine in proteins. *Methods Enzymol* **99**, 387–402.
3. Lodge, A. J., Anderson, J. J., Gullick, W. J., Haugk, B., Leonard, R. C., and Angus, B. (2003) Type 1 growth factor receptor expression in node positive breast cancer: adverse prognostic significance of c-erbB-4. *J Clin Pathol* **56**, 300–304.
4. Machida, K., Mayer, B. J., and Nollau, P. (2003) Profiling the global tyrosine phosphorylation state. *Mol Cell Proteomics* **2**, 215–233.
5. Songyang, Z., Carraway, K. L., III, Eck, M. J., Harrison, S. C., Feldman, R. A., Mohammadi, M., Schlessinger, J., Hubbard, S. R., Smith, D. P., Eng, C., and et al. (1995) Catalytic specificity of protein-tyrosine kinases is critical for selective signalling. *Nature* **373**, 536–539.
6. Pawson, T., Gish, G. D., and Nash, P. (2001) SH2 domains, interaction modules and cellular wiring. *Trends Cell Biol* **11**, 504–511.
7. Schlessinger, J., and Lemmon, M. A. (2003) SH2 and PTB domains in tyrosine kinase signaling. *Sci STKE* **2003**, RE12.
8. Yaffe, M. B. (2002) Phosphotyrosine-binding domains in signal transduction. *Nat Rev Mol Cell Biol* **3**, 177–186.
9. Pawson, T., and Nash, P. (2003) Assembly of cell regulatory systems through protein interaction domains. *Science* **300**, 445–452.
10. Ficarro, S. B., McCleland, M. L., Stukenberg, P. T., Burke, D. J., Ross, M. M., Shabanowitz, J., Hunt, D. F., and White, F. M. (2002) Phosphoproteome analysis by mass spectrometry and its application to Saccharomyces cerevisiae. *Nat Biotechnol* **20**, 301–305.

# 11

# Immunoregulomics

## *A Serum Autoantibody-Based Platform for Transcription Factor Profiling*

Oliver W. Tassinari, Margarita Aponte, Robert J. Caiazzo Jr., and Brian C.-S. Liu

### Summary

Gene expression is regulated by a group of proteins known as transcription factors (TFs). By binding to specific DNA *cis*-elements, each TF contributes a different functional role in gene expression. Panomics™ has developed a TranSignal™ TF–TF Interaction Array, which enables the user to determine TF complexes of interest with multiple other TFs. The process works by immunoprecipitating *cis*-elements bound to native cell nuclear extract TFs using specific antibodies to the TFs, and hybridizing the corresponding *cis*-elements to a membrane array spotted with different consensus sequences. In this protocol, we adapt this technology to characterize and compare autoantibody reactivity to TFs between patients with and without disease. Using Panomics™ combination DNA/protein arrays with over 300 different *cis*-elements spotted on the membrane, we can monitor the differences in autoimmune-targeted regulatory TFs, a process we termed immunoregulomics. This method allows for a qualitative analysis of the interactions with some quantifiable data. The findings can then be verified with the use of gel-shift experiments.

**Key Words:** Autoantibodies; interactome microarrays; transcription factors.

## 1. Introduction

It has been widely established that there is an association between autoimmunity and cancer, as well as well-known autoimmune diseases such as rheumatoid arthritis, multiple sclerosis, and insulin-dependent diabetes mellitus. The identification of the autoantigens targeted by the immune system is

From: *Methods in Molecular Biology, vol. 441: Tissue Proteomics: Pathways, Biomarkers, and Drug Discovery*
Edited by: B.C.-S. Liu © Humana Press, Totowa, NJ

important because the knowledge of these potential autoantigens might better enable a greater understanding of the pathobiology of diseases, facilitate the development of antigen-specific therapies, and aid in the development of potential biomarkers *(1–5)*. Although there have been numerous studies describing the autoimmunity to disease-associated intracellular antigens, few have reported on the autoimmunity to regulatory elements such as transcription factors (TFs).

TFs are a group of proteins responsible for the regulation of gene expression. TFs bind to specific DNA *cis*-elements and essentially "turn on" or "turn off" certain gene expressions. The interactions of TFs are important in understanding gene expression and intra-cellular communications *(6)*. Consequently, there are several techniques designed to observe the interactions of TFs. Immunoprecipitation and super-gel shift techniques are useful in observing the physical interaction of TFs but are limited in their ability to study few TFs at a time. Recently, DNA array technology has entered the TF arena with the creation of consensus elements that can be arrayed on a solid support such as membranes *(6)*. These arrays can then be used to detect labeled *cis*-elements that have been immunoprecipitated from DNA/protein complexes by antibodies for TFs. The TranSignal™ arrays by Panomics™ were designed to use monoclonal antibodies to detect the TF of interest. In this protocol, we have adapted this technology to characterize and compare autoantibody reactivity to TFs between patients with and without disease by using purified IgGs from patient samples as our source of antibody. With this approach, it is possible to identify the TFs targeted by the autoantibodies of patients with specific disease states, an approach we termed immunoregulomics.

In the first step of this immunoregulomics approach, native nuclear extract containing TFs from cells that are being studied (e.g., prostate cancer) is incubated with biotin-labeled double-stranded oligonucleotide *cis*-element probes (TranSignal™ TF Probe Mix). During this incubation, the native TF proteins bind to their corresponding *cis*-elements and create TF/*cis*-element (protein/DNA) complexes. These complexes are then incubated with serum-purified IgGs from test patients (e.g., prostate cancer patients) and controls. The DNA/protein complexes that react with the IgGs are then immunoprecipitated using protein G-coupled magnetic beads. The *cis*-elements are then eluted and hybridized to an array membrane spotted with over 300 corresponding consensus sequences. By incubating probes immunoprecipitated by test and control sample antibodies on separate arrays, we are able to identify the differential autoimmunity on the regulatory elements. By comparing autoantibodies from control serum and from diseased patients, on parallel arrays, it is possible to discount autoimmunity to TFs that are shared by both samples (*see* **Fig. 1**).

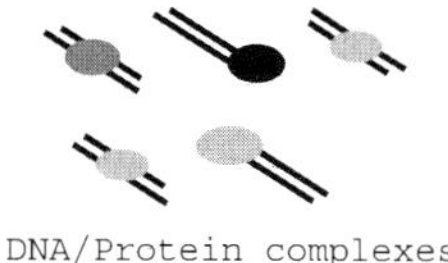


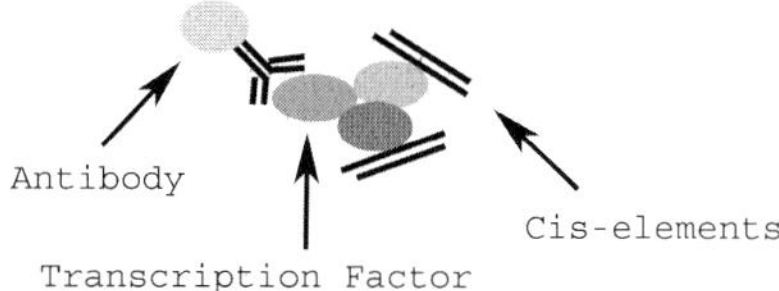


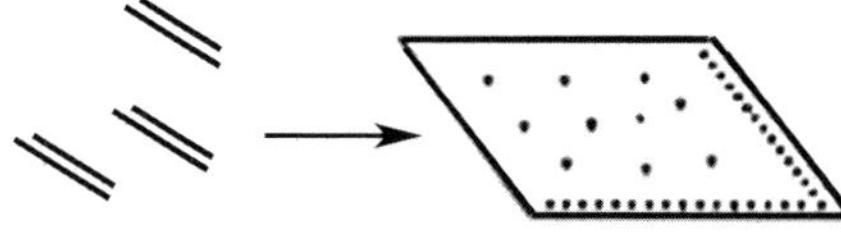


Fig. 1. Schematic chart of TranSignal™ TF Regulomics Array procedure.

In this chapter, we used prostate cancer as a "proof-of-principle" model for our immunoregulomic array approach. Thus, our nuclear extract was derived from cultured prostate tumor cell lines. However, using other sources for nuclear extracts, including whole tissues, this method can be readily adapted to profile the autoantibody reactivity to TFs in other diseases.

## 2. Materials

### 2.1. Nuclear Extraction from Native Cells

1. One 150-mm cell culture plate with 75–90% confluent cells adhered (*see* **Note 1**).
2. Nuclear Extraction Kit (Cat. No. AY2002, Panomics™) containing 10× Buffer A Stock, 5× Buffer B Stock, 10% IGEPAL, 100 nM DTT, Protease Inhibitor Cocktail, and 20× Phosphate-Buffered Saline Stock.
3. Pipettes.
4. Sterile pipette tips.
5. Vortex.
6. Sterile cell lifter or scraper.
7. Sterile 1.5-ml microcentrifuge tubes.
8. Rocking Platform.
9. BCA™ Protein Assay Reagent Kit (Cat. No. 23225, Pierce Biotechnology, Rockford, Illinois, USA) containing BCA™ Reagent A., BCA™ Reagent B, and Albumin standard ampules (2 mg/ml).

## *2.2. Purification of Patient Serum IgG*

1. Serum samples acquired in accordance with IRB-approved protocol and stored at $-80\,^{\circ}$ C.
2. Pipettes.
3. Pipette tips.
4. Microcentirifuge.
5. Vortex.
6. Melon™ Gel Kit (Cat. No. 45206, Pierce Biotechnology) containing Melon™ Gel IgG Purification Support, Melon™ Gel Purification Buffer, and Handee™ Mini-Spin Columns and accessories.
7. End-over-end rotator.
8. BCA™ Protein Assay Reagent Kit (Cat. No. 23225, Pierce Biotechnology) containing BCA™ Reagent A, BCA™ Reagent B, and Albumin standard ampules (2 mg/ml).
9. 96-well plate or other size-appropriate plate for spectrometer.
10. Spectrometer capable of reading absorbance at 562 nm.

## *2.3. Preparation of TF-bound DNA, Immunoprecipitation, and Elution*

1. Panomics™ TranSignal™ TF–TF Interaction Array Reaction Kit (Cat. No. MA5011, Panomics, Inc., Box 2) containing TranSignal™ Probe Mix, Poly d(I-C), 2× Blocking Buffer, 5× Binding Buffer, 5× IP Wash Buffer, 1× IP Dilution Buffer, 1× IP Elution Buffer.
2. Distilled $H_2O$ (RNase, DNase-free).
3. Sterile 1.5-ml microcentrifuge tubes.
4. Pipettes.
5. Pipette tips.
6. End-over-end rotator.
7. Vortex.
8. Magnetic separation stand.
9. Dynabeads™ (Prod. No. 100.03, Invitrogen, Carlsbad, California, USA) with Protein G.
10. Centrifuge.
11. Heating block or water bath.

## *2.4. Hybridization and Detection*

1. Panomics TranSignal™ TF–TF Interaction Array Membranes and Hybridization Reagents (Box 1) containing TranSignal™ TF–TF Interaction Arrays, Hybridization Buffer, Substrate Solution I, Substrate Solution II, Substrate Solution III, 1000× Streptavidin HRP Conjugate, 20× SSC, 20%SDS, 4× Wash Buffer, 10× Detection Buffer, Distilled $H_2O$ (RNase, DNase-free).
2. Hybridization Wash Buffer I: distilled $H_2O$ (RNase, DNase free), 20% SDS, 20× SSC. These reagents are found in the Panomics TranSignal™ TF–TF Interaction Array Membranes and Hybridization Reagents (Box 1).

3. Hybridization Wash Buffer II: distilled $H_2O$ (RNase, DNase free), 20% SDS, 20X SSC. These reagents found in the Panomics TranSignal™ TF–TF Interaction Array Membranes and Hybridization Reagents (Box 1).
4. Hybridization bottles.
5. Hybridization oven.
6. Rocking platform.
7. Water bath.
8. Pipettes.
9. Pipette tips.
10. Plastic wrap.
11. Hyperfilm ECL.
12. Chemiluminescence imaging system.

## 3. Methods

### *3.1. Nuclear Extraction from Native Tissue*

1. All kit solutions should be thawed and kept on ice at all times.
2. Wash cell culture plate with 10 ml of cold 1× PBS twice (*see* **Notes 1** and **2**).
3. Prepare Buffer A Mix in a clean, sterile 1.5 ml-microcentrifuge tube with 1 ml of 1× Buffer A, 10 µl of 100 mM DTT, 10 µl of Protease Inhibitor Cocktail, and 40 µl of 10% IGEPAL.
4. Add 1.0 ml of Buffer A mix to the cell culture plate. Put the plate on ice and rock for 10 min.
5. Scrape cells with a sterile scraper or lifter and pipette into a sterile 1.5-ml microcentrifuge tube. Make sure to pipette up and down a few times as to disrupt the cell clumps.
6. Centrifuge at maximum speed (15,000 × *g*) for 3 min at 4 °C.
7. Place tubes on ice.
8. Remove supernatant and discard.
9. Prepare Buffer B Mix in a clean, sterile 1.5-ml microcentrifuge tube with 147 µl of 1× Buffer B, 1.5 µl of protease inhibitor, and 1.5 µl 100 mM DTT.
10. Resuspend pellet in 150 µl of Buffer B Mix by vortexing.
11. Lay the tube on ice and rock for 2 h.
12. Centrifuge at maximum speed (15,000 × *g*) for 5 min at 4 °C. Collect the supernatant—this is your nuclear extract. Aliquot 5 µl of each sample and measure the protein concentration using the BCA™ Protein Assay Kit. Store at –80 °C until use.

### *3.2. Purification of Patient Serum IgG*

The following steps may be carried out several days before the rest of the experiment. Purified IgG can be stored for up to 1 week at 4 °C. If IgG is to be stored for longer than 1 week, aliquots should be placed in a –20 °C freezer for storage until use; avoid repeated freeze/thaw cycles.

1. Equilibrate the Melon™ Gel IgG Purification Support and Purification Buffer to room temperature (about 30 min) and swirl the bottle containing the Purification Support (do not vortex) to obtain an even suspension. To ensure proper gel slurry while dispensing, use a wide bore or cut pipette tip to dispense 500 µl of gel slurry into a Handee™ Mini-Spin Column placed in a microcentrifuge tube. Swirl the bottle of gel slurry before pipeting each sample to maintain an even gel suspension.

2. Centrifuge the uncapped column/tube assemblies for 1 min at 3000 × *g*, then remove the spin columns and discard the flow-through.

3. Add 300 µl of Purification Buffer to the column, pulse the centrifuge for 10 s and discard the flow-through. Repeat this wash twice. Place the bottom caps on the columns.

4. Add 50 µl of each serum sample diluted 1:10 in 1× Melon™ Gel Purification Buffer to a column. Cap the columns and incubate for 5 min at room temperature with end-over-end rotation.

5. Remove the bottom caps from the columns, loosen the top cap, and insert the spin columns into fresh 2-ml collection tubes. Then, centrifuge for 1 min at 3000 × *g* to collect the purified antibody in the collection tubes.

6. Set up a new column corresponding to each sample that has been purified and repeat **steps 2–5** in order to further purify the collected IgG using fresh Melon™ Gel.

7. Measure the concentration of IgG in each purified sample using Pierce's BCA™ Protein Assay Reagent Kit.

8. Dilute each sample to 1 mg antibody/ml by adding the appropriate volume of 1× Melon™ Gel Purification Buffer. The final amount of IgG should be at least 100 µg per sample.

### 3.3. Preparation of TF-Bound DNA, Immunoprecipitation, and Elution

1. In a sterile 1.5-ml microcentrifuge tube combine 30–100 µg of Nuclear Extract, 10 µl of Poly D (I-C), 10 µl of TranSignal™ TF-TF Probe, 15 µl of 5× binding buffer, and 30 µl of Distilled $H_2O$ (RNase, DNase-free) (*see* **Note 2**).

2. Mix well by pipeting.

3. Incubate samples at 15 °C for 30 min. Afterwards, incubate on ice for an additional 30 min.

4. To the TF-Bound DNA complex created in **step 3**, add 10 µl (this amount should be approximately 10 µg, although the amount of antibody used is completely variable and up to the user) of purified serum patient IgG, as well as 200 µl of 1× dilution buffer.

5. Mix well by gently tapping the tube. Place in end-over-end rotator at 4 °C for 90 min.

6. Before the antibody incubation is complete, prepare the Dynabeads™ by gently shaking the container to resuspend the beads. Pipette 75 µl of the Dynabeads™ solution into a new, sterile, labeled 1.5-ml microcentrifuge tube. Place tube into the magnetic separation stand for 1 min and then pipette off and discard supernatant. Equilibrate the beads by resuspending them with 200 µl 1× dilution

buffer. Again, place the tube into the magnetic separation stand for 1 min and pipette off and discard the supernatant.

7. When incubation with the antibody is complete, place the DNA/TF/Antibody complex into the appropriately labeled tube containing the equilibrated Dynabeads™. Resuspend the beads and incubate with end-over-end rotating for 1 h at 4 °C.

8. If you are proceeding directly to the overnight hybridization of the probe, this is a good place to begin preparing for the hybridization of the array membrane with hybridization buffer (*see* **Subheading 3.4** and **Note 3).**

9. When the 1-h incubation is complete, briefly spin the tube in a microcentrifuge at $400 \times g$ for 2 s and then place the tube into the magnetic separation stand for 2 min to collect the beads. Keeping the tubes on the magnetic separation stand, pipette and discard the supernatant.

10. Wash the beads by adding 400 µl of pre-chilled 1× IP Wash buffer to the tube containing the beads. Completely re-suspend the beads by gently tapping the tube, gently invert the tube four times to clean the sides of the tube, spin the tube in a microcentrifuge at $400 \times g$ for 2 s and then place the tube into the magnetic separation stand for 2 min to collect the beads. Pipette off the fluid and discard. Important: Be sure to pipette off as much fluid as possible in this step. Repeat this step three times.

11. Fill a beaker with water and heat to 100 °C (boiling) to prepare for **step 14**.

12. Prepare the 1× IP Elution Buffer by warming it long enough to dissolve any precipitates before use.

13. Add 60 µl of 1× IP Elution Buffer to the tube containing the beads. Gently mix.

14. Incubate the tube at 100 °C for 5 min (*see* **Note 4**).

15. Transfer to ice immediately. Keep on ice for 2 min.

16. Spin tube in microcentrifuge at $400 \times g$ for 5 s.

17. Place magnetic stand on ice, and place the tube into the stand for 2 min to collect the beads.

18. Transfer the fluid to a clean microcentrifuge tube and place the tube on ice. This is your eluted probe. The beads can be discarded. If you do not plan to use the probe immediately, store it at –20 °C. Before use, thaw the probe, and repeat **steps 14** and **15** to ensure complete denaturation of the double-stranded DNA (*see* **Note 5**).

## *3.4. Hybridization and Detection*

1. Place each array membrane into a hybridization bottle. To wet the membranes, fill each hybridization bottle with 50 ml distilled $H_2O$ (DNase, RNase free). Then, carefully decant the water.

2. To each bottle containing an array membrane, add 3–5 ml of pre-warmed Hybridization Buffer (provided). Place each bottle into a rotating hybridization oven, preheated to 42 °C, for 2 h.

3. Add the two eluted probes to their respective hybridization bottles and hybridize the probes with the membranes at 42 °C overnight.

4. Before finishing the overnight hybridization, prepare Hybridization Wash I and Hybridization Wash II for use in **step 6** (Hybridization Wash Buffer I: 262.5 ml Distilled Water, 7.5 ml 20% SDS, 30 ml 20× SSC. Hybridization Wash Buffer II: 291 ml Distilled Water, 7.5 ml 20% SDS, 1.5 ml 20× SSC). Pre-warm each Hybridization Buffer in a water bath at 35–50 °C.

5. Decant the hybridization mixture from each hybridization bottle.

6. To wash each membrane, add 50 ml of pre-warmed Hybridization Wash I and incubate at 48 °C for 20 min in the rotating hybridization oven. Decant the liquid and repeat the wash. Decant the second wash and add 50 ml of pre-warmed Hybridization Wash II. Again, incubate at 48 °C for 20 min in the rotating hybridization oven. Decant the liquid and repeat the wash.

7. Using forceps, carefully remove each membrane from its hybridization bottle and transfer it face up to a new container containing 20 ml of 1× Blocking Buffer; each membrane needs its own container (we use a container that is equivalent to the size of a 200-ml pipette-tip container, ∼4.5 × 3.5 inches).

8. Incubate the arrays in the 1× Blocking Buffer at room temperature for 30 min with gentle rocking.

9. Transfer 1 ml of Blocking Buffer from each membrane container to a fresh 1.5-ml microcentrifuge tube. To each tube, add 20 µl of 1000× streptavidin–HRP Conjugate and mix well. Return each mixture to the appropriate membrane container and incubate at room temperature for 20 min with rocking.

10. Decant the Blocking Solution containing the antibody. Wash each membrane three times at room temperature with 20 ml of 1× Wash Buffer, with each wash lasting for 5 min.

11. Add 20 ml of 1× Detection Buffer to each membrane and incubate for 5 min at room temperature.

12. To prepare the substrate solution, mix 1 ml of Substrate Solution I and 1 ml of Substrate Solution II. Add 1 ml Substrate Solution III, and mix well.

13. Place each membrane onto a plastic sheet (plastic wrap or a transparent overhead projector sheet will do). Then, pipette 3 ml of substrate solution onto each membrane and overlay each with a second plastic sheet. Ensure that the substrate is evenly distributed over the membrane, with no air bubbles. Incubate at room temperature for 5 min (*see* **Note 3**).

14. Remove excess substrate by gently applying pressure over the top sheet. Using a paper towel, remove excess substrate that might be remaining on the surface of the sheets. Expose the membranes using either Hyperfilm ECL or chemiluminescence imaging system, such as the FluorChem imager from Alpha Innotech Corp. In either case, we recommend that you try several different exposures of varying lengths of time (e.g., 1–4 min). Overexposure of your blot may result in excessive background.

## *3.5. Results and Analysis*

1. Visibly dark spots should be taken into account. If using a control array, compare the images of the test and control arrays and notice the differences. Spots on the

test array that are in an area where there are no spots on the control array should be noted as significant differences and vice versa (for example, if cancer patient serum antibody was used on one array and normal patient serum antibody was used on another, it can be inferred that spots that appear on the test array that do not appear on control array may be possible tumor-related TF) (*see* **Fig. 2**).
2. As a way of quantifying the results of the experiment, the array images can be scanned and programs capable of calculating the density of a given spot can be used to create a raw data number for each spot on the array. Normalization of

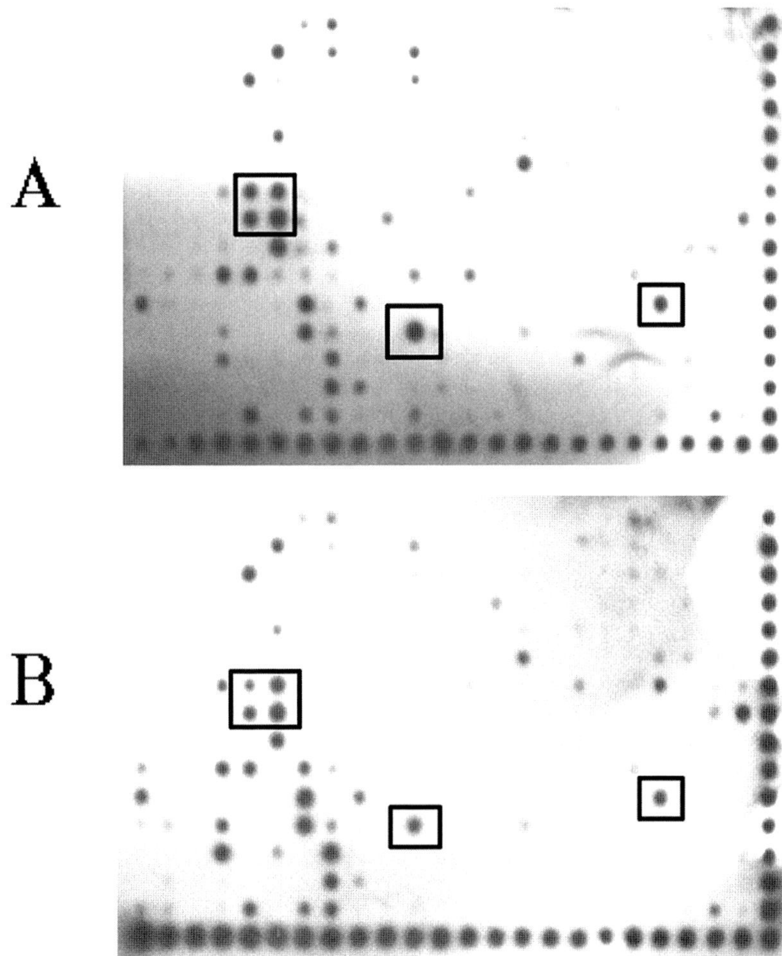

Fig. 2. Two test arrays using prostate cancer patient serum antibody with nuclear extract samples from the same type of cells (LNCaP). The two different hybridization images show the consistency of our approach.

raw number data can be acquired using basic statistical analysis techniques. As previously mentioned, chemiluminescent imaging systems such as FlourChem are available that will both acquire the images and supply software capable of measuring the density of each spot.

3. TFs in question can then be verified using the EMSA gel-shift technique.

## 4. Notes

1. Because, in this chapter, we focus on using cell culture as the source of our nuclear extract, it should be noted that if using whole tissue instead, the proper protocol should be followed. For example, if using the Panomics™ Nuclear Extraction Kit, follow the specified protocol for starting with whole tissue.

2. It should be noted that sterility is very important when performing the nuclear extraction and first few steps of the TF protocol. Make sure that the water that is being used is RNase-free, DNase-free, and sterile. All centrifuge tubes must be sterile or previously autoclaved. All pipeting and transferring of materials should take place under a sterile hood. The derived nuclear extract is very sensitive to post extraction modifications and should therefore be handled very carefully. Once a sample is extracted and the protein concentration is taken, the sample should be used immediately or else aliquoted and frozen at $-80\,^{\circ}\mathrm{C}$. Also, in order to assure consistency of results between trials, it is beneficial to use aliquots from the same original batch.

3. There are several products that can be used as chemiluminescent substrates for the detection of HRP. Substrates I–III that come with the Panomics™ Box 1 are suggested for use in the *Methods* section. Other chemiluminescent substrate kits (such as SuperSignalR West Pico Chemiluminescent substrate by Pierce™ Cat. No. 34080) come in a variety of different intensities. The proper procedure should be followed for use of each substrate kit.

4. The incubation of the elution buffer with the Dynabeads™ can be done on a heat block (as written in the Panomics™ protocol) or in a water bath as well. The use of the hybridization oven is also convenient if it is not being used to hybridize the membranes at the time.

5. Once the probes are eluted, one of two steps can be taken. The probes can be stored in a freezer at $-20\,^{\circ}\mathrm{C}$ until later use or the hybridization of the membrane can be timed so that when the eluted probe is ready it can be placed in the hybridization bottle and oven with the membrane immediately. Timing does not need to be perfect. If the 2-h hybridization preparation period is not over when you are done eluting your probe, the probe can be placed at $-20\,^{\circ}\mathrm{C}$ or on ice until the hybridization is complete.

## Acknowledgments

This work was supported in part by grants U01DK063665 and R01DK066020 from the National Institutes of Health to B.C.S.L.

# References


1. Anderson, K.S. and LaBaer, J. (2005) The sentinel within: Exploiting the immune system for cancer biomarkers. *J Proteome Res* **4**, 1123–1133.
2. Madrid, F.F. (2005) Autoantibodies in breast cancer sera: candidate biomarkers and reporters of tumorigenesis. *Cancer Lett* **230**, 187–198.
3. Arbuckle, M.R., McClain, M.T., Rubertone, M.V., Scofield, R.H., Dennis, G.J., James, J.A., and Harley, J.B. (2006) Development of autoantibodies before the clinical onset of systemic lupus erythematosus. *N Engl J Med* **349**, 1526–1533.
4. Ehrlich, J.R., Qin S., and Liu, B.C.-S. (2006) The "reverse capture" autoantibody microarray: A native antigen-based platform for autoantibody profiling. *Nature Protocols* **1**, 452–460.
5. Qin, S., Qiu, W., Ehrlich, J.R., Ferdinand, A.S., Richie, J.P., O'Leary, M.P., Lee, M.L.T., and Liu, B.C.-S. (2006). Development of a "reverse capture" autoantibody microarray for studies of antigen-autoantibody profiling. *Proteomics* **6**, 3199–3209.
6. Mukhopadhyay, N.K., Ferdinand, A.S., Mukhopadhyay, L., Cinar, B., Lutchman, M., Richie, J.P., Freeman, M.R., and Liu, B.C.-S. (2006) Unraveling androgen receptor interactomes by an array-based method: discovery of proto-oncoprotein c-Rel as a negative regulator of androgen receptor. *Exp Cell Res.* **312**, 3782–3795.

# 12

# The "Reverse Capture" Autoantibody Microarray

*An Innovative Approach to Profiling the Autoantibody Response to Tissue-Derived Native Antigens*

Joshua R. Ehrlich, Liangdan Tang, Robert J. Caiazzo Jr., Daniel W. Cramer, Shu-Kay Ng, Shu-Wing Ng, and Brian C.-S. Liu

## Summary

Recently, we reported the development and use of a "reverse capture" antibody microarray for the purpose of investigating antigen-autoantibody profiling. This platform was developed to allow researchers to characterize and compare the autoantibody profiles of normal and diseased patients. Our "reverse capture" protocol is based on the dual-antibody sandwich immunoassay of enzyme-linked immunosorbent assay (ELISA), and we have previously reported its use to detect autoimmunity to epitopes found on native antigens derived from tumor cell lines. In this protocol, we used ovarian cancer as a model system to adapt the "reverse capture" procedure for use with native antigens derived from frozen tissue samples. The use of this platform in studies of autoimmunity is valuable because it allows for the detection of autoantibody reactivity with epitopes found on the post-translational modifications (PTMs) of native antigens, a feature not present with other protein array platforms. In the first step in the "reverse capture" process, tissue-derived native antigens are immobilized onto the 500 monoclonal antibodies that are spotted in duplicate on the array surface. Using the captured antigens as "baits," we then incubate the array with labeled IgG from test and control samples, and perform a two-slide dye-swap to account for any dye effects. Here, we present a detailed description of the "reverse capture" autoantibody microarray for use with tissue-derived native antigens.

**Key Words:** Autoantibodies; autoantigens; protein microarrays; biomarkers; tissues; ovarian cancer.

## 1. Introduction

Autoimmunity is a key feature of many disease states. Recently, the association of autoimmunity with cancer has been well established *(1–3).*

From: *Methods in Molecular Biology, vol. 441: Tissue Proteomics: Pathways, Biomarkers, and Drug Discovery*
Edited by: B.C.-S. Liu © Humana Press, Totowa, NJ

In cancer, host antibodies react with a unique group of autologous intracellular proteins known as tumor-associated antigens (TAAs). Additionally, autoimmunity is consistently observed in a variety of well-known autoimmune diseases, including rheumatoid arthritis, multiple sclerosis, and insulin-dependent diabetes mellitus. The detection of autoimmunity is useful because autoantibodies can serve as biomarkers for disease, and their presence may help to elucidate the role of significant disease-related biochemical pathways.

Previous studies of autoimmunity using microarray technology have relied on recombinant proteins and/or synthetic peptides as arrayed features *(4–7)*. The major shortcoming of this approach is that these platforms may fail to detect the full range of autoimmunity present in a given disease state because of the absence of native protein conformations and relevant post-translational modifications (PTMs) of the arrayed features. Our laboratory recently reported on the development of a "reverse capture" autoantibody microarray *(8,9)*. The "reverse capture" approach overcomes this significant shortcoming of other array platforms by detecting autoimmunity to proteins derived from a native source such as frozen tissue samples.

The "reverse capture" autoantibody microarray is based on the dual-antibody sandwich immunoassay of enzyme-linked immunosorbent assay (ELISA) (*see* **Fig. 1**). The basic platform consists of a glass microarray slide arrayed with 1000 highly specific well-characterized monoclonal antibodies against 500 unique antigens. These antibodies are used to immobilize native proteins. Because the reagents used in the procedure are non-denaturing, antigens are

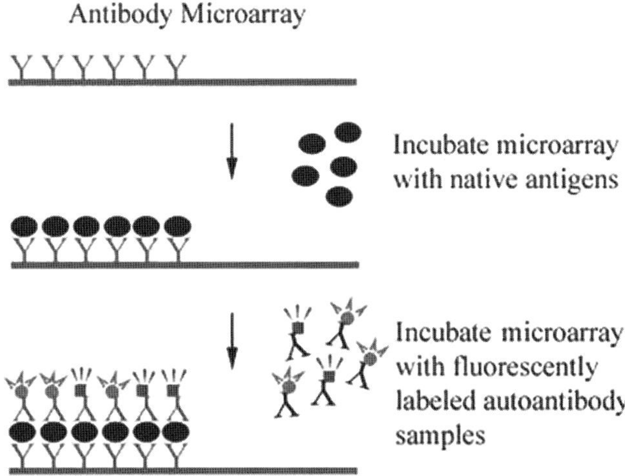


Fig. 1. A schematic representation of the "reverse capture" autoantibody microarray platform.

immobilized on the array in their native conformations with all of their appropriate PTMs. The antigens captured on the slide then serve as "baits" for differentially labeled patient IgG samples. By labeling test and control IgG samples with different fluorescent dyes and scanning the "reverse capture" slides, it is possible to identify the target antigens to which a test sample differentially expresses autoantibodies, as compared with the control. We term this array platform a "reverse capture" autoantibody microarray, because in standard microarray nomenclature, "capture" refers to the immobilization and detection of the analyte-antigen, not the antibody.

Additionally, the "reverse capture" protocol employs a two-slide dye-swap method in which the dye/sample pairings used on one microarray slide are reversed and applied to a second, parallel slide. This component of the protocol facilitates the normalization of data, and controls for differences in the labeling efficiencies of the different dyes, as well as differences in antibody-binding efficiencies following the labeling reactions.

We have determined several key factors which must be accounted for in the analysis of "reverse capture" data: (a) variation in average spot fluorescence intensity between dye channels, (b) outlying/aberrant data points, and (c) average background intensity. Accordingly, we suggest eliminating far-outlying data points and centering all data, through the method of ratio-based normalization, before determining significant autoantibody expression using a Student's *t*-test and hierarchical clustering. There are numerous commercially available computer programs that will assist with these types of analyses, as well as with the construction of heat maps for graphical presentation of your results. In this chapter, we describe a general normalization, transformation, and analysis method that can be carried out using any high-quality software package. We also introduce the recently developed Mixed-effects Model that can be used to account for interdependence and correlation of observations.

Additional information regarding our biostatistical methods and findings using the "reverse capture" platform is available *(8,10)*. In this chapter, we used ovarian cancer as a model system to adapt the "reverse capture" platform for use with native antigens derived from frozen tissue samples. This method, however, may be used to study differential autoantibody expression in any disease where autoimmunity is presumed to exist.


## 2. Materials

### *2.1. Purification of IgG*

1. Serum samples acquired in accordance with IRB-approved protocol and stored at −80 °C.
2. Pipettes.

3.  Pipette tips.
4.  Microcentrifuge.
5.  Vortex.
6.  Melon™ Gel Kit (Cat. No. 45206, Pierce Biotechnology, Rockford, IL, USA) containing Melon™ Gel IgG Purification Support, Melon™ Gel Purification Buffer, and Handee™ Mini-Spin Columns and accessories.
7.  End-over-end rotator.
8.  BCA™ Protein Assay Reagent Kit (Cat. No. 23225, Pierce Biotechnology) containing BCA™ Reagent A., BCA™ Reagent B, and Albumin standard ampules (2 mg/ml).
9.  96-well plate or other size-appropriate plate for spectrometer.
10. Spectrometer capable of reading absorbance at 562 nm.

### *2.2. Labeling of IgG with Fluorescent Dyes*

1.  DyLight™ 547 and Dylight™ 647 Monoclonal Antibody Labeling Kits (Cat. Nos. 53009 and 53015, Pierce Biotechnology) containing DyLight™ 547 and 647 NHS Esters, dimethylformamide (DMF), Borate buffer (0.67 M), and Zebra™ Desalt Spin Columns.
2.  0.5-ml microfuge tubes.
3.  Pipettes.
4.  Pipette tips.
5.  Vortex.
6.  Microcentrifuge.

### *2.3. Removal of Unbound Dye*

1.  Zebra™ Desalt Spin Columns (Cat. No. 89882, Pierce Biotechnology, also supplied as a component of the Pierce Dylight™ Monoclonal Antibody Labeling Kits).
2.  Kimwipes.
3.  Pipettes.
4.  Pipette tips.
5.  2-ml microcentrifuge collection tubes.
6.  Microcentrifuge.
7.  Ultra-pure water.

### *2.4. Native Protein Extraction from Tissue*

1.  50–100 mg of frozen tissue.
2.  Mortar and pestle.
3.  Alumina (Cat. No. A-2039, Sigma, St. Louis, MO, USA).
4.  Ab Microarray Buffer Kit (Cat. No. 631792, Clontech, Mountain View, CA, USA) containing, Incubation Tray, Extraction/Labeling Buffer, Stock Incubation Buffer, Background Reducer, and Wash Buffers A–C.

   5.  Microcentrifuge.
   6.  1.5- or 2-ml microcentrifuge tubes.
   7.  Pipettes.
   8.  Pipette tips.
   9.  BCA™ Protein Assay Reagent Kit (Cat. No. 23225, Pierce Biotechnology).
10.  96-well plate or other size-appropriate plate for spectrometer.
11.  Spectrometer capable of reading absorbance at 562 nm.

### 2.5. Antibody Array Incubation with Native Antigens

1. Ab Microarrays (Cat. No. 631790, Clontech) containing 2 Ab Microarrays, and one Storage Vial (empty).
2. Ab Microarray Buffer Kit (Cat. No. 631792, Clontech).
3. 50-ml conical centrifuge tubes.
4. Rocking platform.
5. Pipettes.
6. Pipette tips.
7. Phosphate-buffered saline (PBS) (pH 7.4).

### 2.6. Antibody Microarray Incubation with Patient IgG

1. Ab Microarray Buffer Kit (Cat. No. 631792, BD Clontech).
2. 15-ml conical centrifuge tubes.
3. Forceps.
4. Pipettes.
5. Pipette tips.
6. Rocking platform.
7. Kimwipes.
8 Swinging bucket centrifuge with adaptor for 50-ml tubes.

### 2.7. Microarray Scanning and Statistical Analysis

1. Microarray scanner compatible with $75 \times 25 \times 1$-mm slides and capable of dual-color analysis. The scanner must be capable of measuring fluorescence in the ranges of the Cy3 and Cy5 fluorescent labels.
2. Microarray scanning software.
3. Microsoft Excel.
4. Microarray data analysis software.

## 3. Methods

The following protocol has been optimized to detect the presence of autoanti-bodies directed against native host antigens derived from frozen tissue samples.

This procedure should be carried out at room temperature wearing latex or nitrile examination gloves, and all reagents used in the protocol should be kept at 4 °C or on ice at all times, unless otherwise specified explicitly in the protocol.

### 3.1. Purification of IgG using Melon™ Gel Kit

The following steps may be carried out several days before the rest of the experiment. If you have not purified IgG before the day of your "reverse capture" experiment, we recommend that you perform the purification and labeling of patient IgG during the Incubation with Native Antigens (*see* **Subheading 3.5.**); this will allow you to optimize the time spent performing this procedure. Purified IgG can be stored for up to 1 week at 4 °C. If IgG is to be stored for longer than 1 week, aliquots should be placed in a -20 °C freezer for storage until use; avoid repeated freeze/thaw cycles.

1. Equilibrate the Melon™ Gel IgG Purification Support and Purification Buffer to room temperature (about 30 min) and swirl the bottle containing the Purification Support (do not vortex) to obtain an even suspension. To ensure proper gel slurry dispensing, use a wide bore or cut pipette tip to dispense 500 μl of gel slurry into a Handee™ Mini-Spin Column placed in a microcentrifuge tube. Swirl the bottle of gel slurry before pipetting each sample to maintain an even gel suspension.
2. Centrifuge the uncapped column/tube assemblies for 1 min at 3000 × *g*, then remove the spin columns and discard the flow-through.
3. Add 300 μl of Purification Buffer to the column, pulse the centrifuge for 10 s, and discard the flow-through. Repeat this wash twice. Place the bottom caps on the columns.
4. Add 50 μl of each serum sample diluted 1:10 in 1× Melon™ Gel Purification Buffer to a column. Cap the columns and incubate for 5 min at room temperature with end-over-end rotation.
5. Remove the bottom caps from the columns, loosen the top cap, and insert the spin columns into fresh 2-ml collection tubes. Then, centrifuge for 1 min at 3000 × *g* to collect the purified antibody in the collection tubes.
6. Set up a new column corresponding to each sample that has been purified and repeat **steps 2–5** in order to further purify the collected IgG using fresh Melon™ Gel (*see* **Note 1**).
7. Measure the concentration of IgG in each purified sample using Pierce's BCA™ Protein Assay Reagent Kit.
8. Dilute each sample to 1 mg antibody/ml by adding the appropriate volume of 1× Melon™ Gel Purification Buffer (*see* **Note 2**). The final amount of IgG should be at least 200 μg per sample.

## 3.2. Labeling of IgG with Fluorescent Dyes

The following steps may be carried out several days before the rest of the experiment. However, **Subheadings 3.2** and **3.3** must be carried out together so that unbound dye is removed before storage of the labeled IgG. Labeled IgG can be stored for a maximum of 1 week at 4 °C protected from light. If labeled IgG is to be stored for longer than 1 week, aliquots should be protected from light in a –20 °C freezer until use. Repeated freeze/thaw cycles should be avoided.

Complete **steps 1–10** rapidly without interruption. Once the DyLight™ dyes are reconstituted, they must be used immediately (*see* **Note 3**).

1. Set up and label one 0.5-ml microfuge tube for each sample (four tubes total: A-DyLight™ 547, A-DyLight™ 647, B-DyLight™ 547, B-DyLight™ 647).
2. Transfer 100 µg of the appropriate purified antibody (1 mg/ml) from **Subheading 3.1** to the corresponding tube prepared in **step 1** and add 10 µl of Borate Buffer (0.67 M) to each tube.
3. Tap the bottom of the DyLight™ Reagent vials against a hard surface to ensure there is no dye in the caps and reconstitute one vial of DyLight™ 547 Reagent and one vial of DyLight™ 647 Reagent by adding 20 µl of DMF to each vial.
4. Vortex the two dye vials and briefly centrifuge to collect the reconstituted dyes at the bottom.
5. Add 8 µl of DyLight™ 547 to each of the corresponding tubes from **step 2**.
6. Add 5 µl of DyLight™ 647 to each of the corresponding tubes from **step 2**.
7. Vortex the four microfuge tubes gently and briefly centrifuge the microfuge tubes to collect the samples at the bottom.
8. Incubate the tubes for 45 min at room temperature protected from light.
9. Proceed immediately with the removal of unbound dye.

## 3.3. Removal of Unbound Dye with Desalting Columns

As long as you work quickly, desalting can be completed at room temperature. Otherwise, if you have access to a cold room, we suggest you complete the procedure at 4 °C.

1. Use two Zebra™ Desalt Spin Columns for each sample (8 columns in total) (*see* **Note 4**).
2. Twist off the bottom of each column and loosen the caps before placing each one in its collection tube and centrifuge each column at 1500 × *g* for 1 min to remove the storage buffer. Note the side of each column where the compacted resin is slanted upwards, and be sure to place the columns in the centrifuge with this area of the column facing outward in all subsequent centrifugations.
3. Blot the bottom of the columns against a laboratory tissue to remove excess liquid.

4. Place each column into a fresh 2-ml collection tube and label two tubes for each sample: A-DyLight™ 547, A-DyLight™ 647, B-DyLight™ 547, B-DyLight™ 647 (8-column/tube assemblies in total).

5. Remove the caps from the columns and carefully apply half ($\sim$60 µl) of each labeling reaction directly onto the center of the resin bed of its corresponding column. Each labeling reaction tube should have two corresponding desalting columns.

6. After the samples have been fully absorbed ($\sim$2 min protected from light), apply 15 µl of ultrapure water to the resin bed of each column and centrifuge the columns for 2 min at 1500 $\times$ $g$ to collect the desalted, labeled antibodies.

7. Combine the paired desalted samples so that each sample, A-DyLight™ 547, A-DyLight™ 647, B-DyLight™ 547, B-DyLight™ 647, is consolidated into one 0.5-ml tube and has a volume of approximately 140 µl.

8. At this point, purified, labeled IgG samples can be stored for future use or used in **Subheading 3.6** of a current experiment.

### 3.4. Extraction of Native Antigen from Frozen Tissue

This protocol employs the Clontech Ab Microarray 500; however, in a different capacity from that set forth by the manufacturer. Nevertheless, in order to use this microarray platform for autoantibody profiling, native antigens should be extracted from a frozen tissue sample in the *exact* same method set forth in Clontech's protocol. Please refer to and follow the protocol in **Subheading 3.1.1** (*Extracting Protein from Tissue Samples*) of Chapter 9, in order to extract native antigens from your frozen tissue sample. After completing the antigen extraction, continue immediately to **Subheading 3.5** (Chapter 12) of our "reverse capture" autoantibody microarray protocol (*do not continue with the Clontech protocol*). Also, please note that we recommend performing the native antigen extraction on the same day as the remaining portions of the "reverse capture" protocol. If proteins are extracted at an earlier time, phosphatase activity, proteolysis, and other proteolytic degradations may occur that will alter the outcome of the experiment.

### 3.5. Antibody Array Incubation with Native Antigens

1. Set up the Incubation Tray provided. On the lid of the Incubation Tray, label the four chambers as follows: Slide 1 Incubation, Slide 1 Wash, Slide 2 Incubation, Slide 2 Wash (*see* **Fig. 2**).

2. Prepare 50 ml Incubation Buffer from 45 ml Stock Incubation Buffer and 5 ml Background Reducer. Then, add 5 ml of Incubation Buffer to the incubation chambers.

3. Transfer 200 µg of protein from **Subheading 3.4** to each of the incubation chambers and incubate the tray at room temperature for 30 min with gentle rocking.

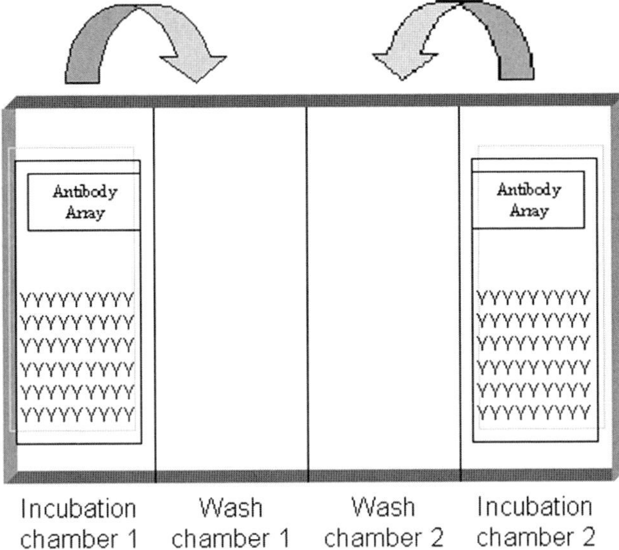



Fig. 2. An illustration of the Incubation Tray showing incubation and wash chambers.

4. About 5 min before the incubation in **step 3** is complete, prepare the Ab Microarrays by washing the slides two times. Wash as follows:

   a. Decant the Storage Buffer from the Storage Vial.
   b. Add 30 ml of Stock Incubation Buffer.
   c. Cap the Storage Vial and slowly invert the vial 10 times.
   d. Decant the Stock Incubation Buffer from the Storage Vial.
   e. Add 20 ml of Incubation Buffer and repeat **steps c** and **d**.

5. Record the lot number of the slides found on the label on the outside of the Storage Vial.
6. Carefully remove the slides from the Storage Vial and place each one, label side facing up, into one of the two designated incubation chambers containing the native antigen mix.
7. Incubate the slides at room temperature for 45 min with constant gentle rocking. Every 15 min, use a fresh pipette tip to gently lift each slide while rocking the incubation tray to allow the native antigen mix to come into contact with all sides of the slide.
8. Remove the buffer from the incubation chambers and discard.
9. Add 5 ml of PBS (pH 7.4) to the incubation chambers, taking care to pipette the liquid on the portion of the slide containing the manufacturer's label.
10. Incubate at room temperature for 5 min with gentle rocking.
11. Repeat **steps 9** and **10** two times.
12. Proceed immediately with the microarray incubation with patient IgG.

### *3.6. Antibody Array Incubation with Patient IgG*

The labeled IgG samples should be incubated with the microarrays in the same incubation chambers used in **Subheading 3.5** for the incubation with native antigens. Some biologically relevant antigens that have not been captured by the arrays' monoclonal antibodies in **Subheading 3.5** will be found on the walls of the incubation chamber. The high affinity of certain IgG molecules for these antigens' epitopes limits non-specific binding of labeled IgG to the microarray slide.

1. Label one 15-ml conical tube "Mix 1" and label one 15-ml conical tube "Mix 2"; then add 5 ml of Incubation Buffer to each of these tubes.
2. Transfer the entire sample of IgG A-DyLight™ 547 and IgG B-DyLight™ 647 (200 µg total; from **Subheading 3.3**) to the Mix 1 tube from **step 1**.
3. Transfer the entire sample of IgG A-DyLight™ 647 and IgG B-DyLight™ 547 (200 µg total; from **Subheading 3.3.**) to the Mix 2 tube from **step 1**.
4. Add the contents of the tube labeled "Mix 1" to incubation chamber 1 and "Mix 2" to incubation chamber 2.
5. Incubate the native protein array slide (prepared in **Subheading 3.5.**) at room temperature for 45 min with gentle rocking. Every 15 min, use a pipette tip to lift one end of the slide while gently rocking the Incubation Tray.
6. Add 5 ml of Incubation Buffer (prepared in **Subheading 3.5.**) to each wash chamber, transfer the slides to their respective wash chambers, and incubate at room temperature for 5 min with gentle rocking.
7. Remove the buffer from the wash chambers.
8. Add 5 ml of Wash Buffer A to each wash chamber.
9. Incubate at room temperature for 5 min with gentle rocking.
10. Repeat **steps 7–9** using Wash Buffer B, and then Wash Buffer C.
11. Dry the slides. It is important to remove as much moisture as possible from the surface of the slides before the liquid evaporates passively.

    a. Using scissors or a knife, puncture a small, round hole in the bottom of the Storage Vial provided. This will facilitate the removal of excess liquid from the slides during centrifugation.
    b. Using gloved hands and touching only the edges of the slides, hold the slides so that the excess liquid drips toward the bottom of the array slides (the area containing the manufacturer's label) and gently touch this edge to a clean Kimwipe several times.
    c. Carefully place the slides in the empty Storage Vial with the ends containing the manufacturer's label at the bottom of the vial. Do not touch the array surface.
    d. Cap the vial and centrifuge the slides at $1000 \times g$ for 25 min at room temperature.

12. Proceed immediately with microarray scanning (*see* **Subheading 3.7.**).

### 3.7. Microarray Scanning and Quantitation

Antibody microarray slides should be scanned using a laser scanner, such as the Axon GenePix 4000B or the Perkin Elmer ScanArray 5000, according to the manufacturer's specifications. The scanner must be able to measure fluorescence in the ranges of the Cy3 and Cy5 fluorophores (*see* **Note 5**).

1. Turn on the scanner and allow the lasers to warm up. The lasers on the Perkin Elmer ScanArray 5000 require 15 min to warm-up before scanning your arrays.
2. Run a quick/preview scan of the entire slide in order to determine the area containing the arrayed features.
3. Create a scan protocol on the computer attached to the microarray scanner (*see* **Note 6**).

    a. Set the protocol to scan for DyLight™ 547 and DyLight™ 647 fluorescence. It is acceptable to use the standard settings for Cy3 and Cy5 in order to scan for the DyLight™ dyes because these fluors have similar excitation/emission maxima (excitation; emission): DyLight™ 547: 557 nm; 570 nm Cy3: 550 nm; 570 nm DyLight™ 647: 652 nm; 673 nm Cy5: 649 nm; 670 nm
    b. Determine the area containing the arrayed monoclonal antibodies and select this as the portion of the array to be scanned.
    c. If possible, select an area on the array that contains no arrayed features to be scanned for background intensity.
    d. Set the laser powers and PMT (photomultiplier) Gains so that the signal is high enough without being saturated (*see* **Note 7**). We suggest the following settings for scanning with the Perkin Elmer ScanArray 5000:
    Cy3/DyLight™ 547: PMT = 62%; laser power = 90%
    Cy5/DyLight™ 647: PMT=50%; laser power=90%

4. Insert the first slide, containing the Mix 1 autoantibodies, into the scanner. For the ScanArray 5000, the microarray surface is inserted face up.
5. Begin the scan and make sure that the signal is sufficiently high, but not saturated. You may need to readjust the laser and PMT Gain settings or rerun an Automatic Sensitivity Calibration if the image is saturated or the signal is too low.
6. Save the DyLight™ 547 and 647 images as separate TIFF files. The single-file TIFF format is the most useful for quantifying data from the images. Other file formats, such as Bitmap (BMP), may be appropriate for other tasks (*see* **Note 8**).
7. Scan the second microarray slide, containing the Mix 2 autoantibodies, using the same settings used to scan the first slide and save the images in the same manner.
8. Obtain the Axon Grid (GAL file) that corresponds to the lot number of the Clontech™ Ab 500 Microarray used. This can be found at http://bioinfo.clontech.com/abinfo/array-list-action.do

9. Using the GenePix Pro software, use the "Alt Y" command to open the GAL file downloaded in **step 8** and open the TIFF files corresponding to your first slide using the "Alt O" command.

10. Automatically align the Axon Grid with the array features using "F8." Use the Zoom-In feature to ensure the proper alignment of the grid with the features of your array. You can adjust the fit of the Grid to the entire array or to individual array features using the tools on the left-hand side of the screen.

11. Carry out an Automatic Analysis using the "Alt A" command (*see* **Note 9**) and save your data as a GPR (GenePix Results) file using the "Alt U" command (*see* **Note 10**).

12. Data can be exported to Excel using the "Ctrl A" command to select all data, followed by the "Ctrl C" command to copy all data to the clipboard.

13. Data from your first slide can now be pasted into a blank Excel sheet with the "Ctrl V" command.

14. Repeat **steps 9–13** using the TIFF images from your second slide (*see* **Note 11**). Note, the same GAL file should be used to extract data from the second slide because slides provided as a pair always have the same lot number.

**Fig. 3** shows the scan images from a "reverse capture" experiment where the dye-swap method was employed. In this experiment, native antigens were extracted from malignant ovarian tissue and hybridized with the two microarray slides, as described. Test IgG was taken from the serum of an ovarian cancer patient, whereas control IgG was prepared from the sera of group of 20 age-matched normal females; IgGs were differentially labeled. The spots that appear both green (DyLight™ 547) on one slide and red (DyLight™ 647) on the other, represent candidate antigen-autoantibody reactivities. The significance of these spots is investigated in **Subheading 3.8**.

### 3.8. Biostatistical Data Analysis

Biostatistical analyses are needed to appropriately normalize, analyze, and interpret the vast amounts of data obtained from "reverse capture" studies. There are many analysis tools to extract reliable information from microarray data [commercial software packages as well as programs provided on the web at no cost to investigators *(11)*]. Regardless of the tools that are employed, "reverse capture" data must be normalized and transformed before reasonable data comparisons can be made (*see* **Note 12**).

1. The raw array data acquired in GenePix Pro have to be pre-processed to separate signals from noise, eliminate low-quality measurements, handle missing values, and adjust the measured intensities to facilitate comparisons *(12,13)*. Image analysis tools reduce the background signal and minimize the impact of poor-quality spots on the final array data *(11,12)*.

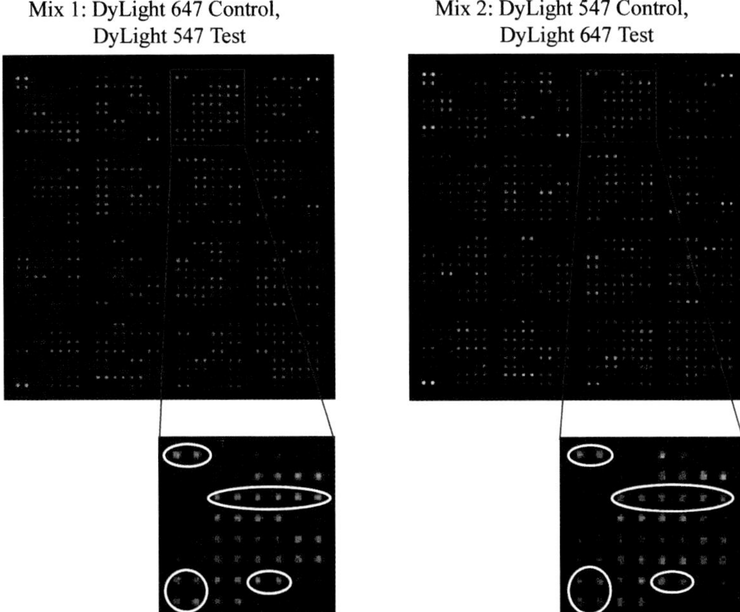



Fig. 3. An illustration of results from the two-slide dye-swap. In Mix 1, control IgG is labeled with DyLight™ 647 dye (red) and ovarian cancer patient IgG is labeled with DyLight™ 547 dye (green). In Mix 2, the dyes are swapped, so control IgG fluoresces green, whereas ovarian cancer IgG fluoresces red. A portion of each array slide is enlarged to clearly illustrate that green spots denoting ovarian cancer autoantibody reactivity in Mix 1, fluoresce red in Mix 2.

2. The simplest method of normalization is a linear transformation of the data to correct for experimental variability because of differences in sample quantity, fluorescent labeling, hybridization efficiencies, and inter-array variations *(14–16)*.

  a. General linear modeling methods such as ANOVA *(17)* or a mixed model *(18)* protocol can be used to link the normalization process with downstream data analysis.

  b. Non-linear normalization approaches, such as locally weighted regression, are useful for removing intensity-dependent effects on the fluorescence ratios *(16,19,20)*.

3. Other data transformation procedures such as intensity-based filtering and replicate filtering can increase the reliability of measurements and assist with downstream clustering. For our dye-swapped replicate data, background-subtracted Log$_2$-transformed intensity ratios of the dye-swapping slides can be plotted against each other; outliers with more than two standard deviations should then be excluded from clustering analysis.

4. Missing value estimation methods based on $k$-nearest neighbors ($k$-NN) or a singular value decomposition (SVD approach) can be used to remedy the data where multiple missing expression values exist *(13,21)*.

5. Following normalization of the data, the underlying patterns of antigen-autoantibody reactivity can be examined by clustering antigen signals into groups with similar patterns *(22,23)*.

   a. Agglomerative hierarchical clustering, $k$-means clustering, and self-organizing maps (SOM) are widely used methods *(23,24)*.

   b. Recently, mixture model-based clustering has become very popular for the clustering analysis of microarray data *(13,25,26)*, as it provides a sound mathematical framework for clustering and a principled statistical approach to the determination of the ideal number of clusters *(13)*.

   c. For clustering correlated antigen-autoantibody reactivities from different kinds of "reverse capture" experiments (such as time–course experiments, experiments with repeated measurements, and cross-sectional experiments), a new method has been proposed through the formulation of a linear mixed-effects model (LMM) *(27)*. This model reduces the possibility of misleading clusters of microarray data by overlooking the presence of correlations between observations and the dependence between microarray data *(28,29)*.

6. For a "supervised" approach to building a classifier suitable for diagnosis, as well as drug discovery, support vector machine (SVM) methods based on a backward selection procedure known as recursive feature elimination (RFE) have been developed to select a small subset of significant data points from broad patterns of data *(30,31)*. Data points are ranked in relative importance as markers on the basis of the absolute values of their fitted coefficients in the linear form of the classifier for an SVM formed with linear kernel *(30)*. In estimating the error rate of the classifier based on a smaller set of significant parameters, care must be taken to ensure that the selection bias associated with choosing the optimal of parameters has been assessed and corrected for, using either an "external" cross-validation or a bootstrap approach *(32)*.

7. Another method for identifying biomarkers from a very large number of candidates is to rank antigen-autoantibody reactivities using a supervised selection procedure that makes use of known clinical data such as survival time, time to distant metastases, or time to recurrence of tumor *(33,34)*.

## 4. Notes

1. The IgG purification procedure is repeated in order to ensure exclusive isolation of IgG from patient sera. We have found that with only one purification run, high-abundance proteins other than IgG may be present in the eluted solution.

2. To label the purified antibodies with fluorescent dyes, the antibodies must be in a buffer free of ammonium ions and primary amines. The Melon™ Gel Purification

Buffer used in **Subheading 3.1** is compatible with all of the DyLight™ labeling reagents used in **Subheading 3.2**.

3. Pierce Biotechnology supplies the DyLight™ 547 and DyLight™ 647 dyes as monofunctional *N*-hydroxysuccinimide (NHS)-esters in dried pre-measured amounts. The NHS-ester is a functional group that cross-links to primary amines. This reaction releases the NHS group and produces a covalent amide bond that links the dye to the amine. Because of the unique affinity ratio of each dye to the antibody that it is labeling, different volumes of the two dyes are used in the labeling reactions. Once reconstituted in the DMF supplied in the DyLight™ labeling kits, unused aliquots of the DyLight™ dyes cannot be saved for future use. The binding efficiency is quickly diminished in this medium. If you wish to save unused portions of the dyes for later use, they should initially be reconstituted in anhydrous DMF (or DMSO) and frozen at –20 °C until use.

4. The Zebra™ Desalt Spin Columns that come with the DyLight™ Labeling Kits remove free dye. These columns contain a desalting resin and molecular weight cutoff. They perform well in desalting small sample volumes, providing excellent protein recovery and $\geq 95\%$ retention of small molecules and salts (<7 kDa).

5. The excitation and emission maxima of the more commonly used CyDye fluorophores are very close to those of the DyLight™ dyes, where DyLight™ 547 is analogous to Cy3 and DyLight™ 647 is analogous to Cy5. Thus, the preprogrammed settings for the scanning of CyDyes may be used in scanning arrays labeled with the Pierce DyLight™ dyes.

6. There are various software packages available for creating scan and quantitation protocols. ScanArray Express by Packard Bioscience or GenePix Pro 6.0 by Molecular Devices is recommended for the scanning of the microarray slides. GenePix Pro is recommended for the quantitation of your results.

7. The microarray may need to be scanned several times before the ideal laser power and PMT settings are ascertained. If your scanner has the option of running an Automatic Sensitivity Calibration, this option may be perused. We recommend that you also refer to our published "reverse capture" troubleshooting tips for additional suggestions *(9)*.

8. If the images will be displayed as graphics in a presentation or article, they should be saved individually as BMP files and then readjusted and/or overlayed in a photo-editing program, such as Adobe Photoshop™.

9. The tab labeled "Histogram" at the top of the screen allows you to check the normalization of your data. When dye and scanning intensities are perfectly normalized, the Count Ratio will be equal to 1.0. If your Count Ratio deviates significantly from 1.0, you should attempt to adjust the PMT gains used for scanning so that this value approximates 1.0. The "Scatter Plot" tab allows you to view the distribution of your data to ensure that there are a minimal number of outlying data points.

10. GPR files are used in some biostatistical analyses. For example, the microarray data analysis software Acuity 4.0 by Molecular Devices uses this file format

in order to perform data manipulation and analyses, including ratio-based normalization, lowess normalization, hierarchical clustering, non-hierarchical clustering, and *t*-tests for statistical significance.

11. When opening the TIFF files from the second slide, it is necessary to check the box next to "Replace current images" in the "Open Images" dialogue box.
12. Normalization and transformation of data accomplish two aims. First, log-transformation of data scales the range of your data; as "reverse capture" data is acquired in ratio form, it is important that the scale allows for equal differential expression by the sample represented in both the numerator and the denominator of the ratio. Second, normalization centers data around an expected mean value; log-transformed data are centered around the expected median (or mean) log-ratio of 0.0 across all data points.

## Acknowledgments

This work was supported, in part, by Grant U01DK063665 from the National Institutes of Health to B.C.-S. Liu. Liangdan Tang is supported by a fellowship from China Scholarship Council (CSC). Shu-Wing Ng is partially supported by a Clinical Innovator Award from the Flight Attendant Medical Research Institute (FAMRI).

## Disclosure

International patent protection is currently pending for the "reverse capture" autoantibody microarray platform.

## References

1. Prehn, R.T. (2006) An adaptive immune reaction may be necessary for cancer development. *Theor. Biol. Med. Model* **3**(1), 6.
2. Dunn, G.P., Old, L.J., and Shreiber, R.D. (2004) The immunobiology of cancer immunosurveillance and immunoediting. *Immunity* **21**,137–148.
3. Madrid, F.F. (2005) Autoantibodies in breast cancer sera: Candidate biomarkers and reporters of tumorigenesis. *Cancer Lett.* **230**,187–198.
4. Mintz, P.J., Kim, J., Do, K.A., Wang, X., Zinner, R.G., Cristofanilli, M., Arap, M.A., Hong, W.K., Troncoso, P., Logothetis, C.J., Pasqualini, R., and Arap, W. (2003) Finger printing the circulating repertoire of antibodies from cancer patients. *Nat Biotechnol.* **21**, 57–63.
5. Wang, X., Yu, J., Sreekumar, A., Varambally, S., Shen, R., Giacherio, D., Mehra, R., Montie, J.E., Pienta, K.J., Sanda, M.G., Kantoff, P.W., Rubin, M.A., Wei, J.T., Ghosh, D., and Chinnaiyan, A.M. (2005) Autoantibody signatures in prostate cancer. *N. Engl. J. Med.* **353**, 1224–1235.

6. Hueber, W. (2005) Antigen microarray profiling in rheumatoid arthritis. *Arthritis Rheum.* **52**, 2645–2655.

7. Robinson, W.H., Steinman, L., and Utz, P.J. (2002) Protein and peptide array analysis of autoimmune disease. *BioTechniques* **33**, S66–S69.

8. Qin, S., Qui, W., Ehrlich, J.R., Ferdinand, A.S., Richie, J.P., O'Leary, M.P., Lee, M.L.T., and Liu, B.C.-S. (2006) Development of a "reverse capture" autoantibody microarray for studies of antigen-autoantibody profiling. *Proteomics* **6**, 3199–3209.

9. Ehrlich, J.R., Qin, S., and Liu, B.C-S. (2006) The 'reverse capture' autoantibody microarray: A native antigen-based platform for autoantibody profiling. *Nat. Protocols* **1**, 452–460.

10. Liu, B.C-S. and Ehrlich, J.R. (2006) Proteomics approaches to urologic diseases. *Expert Rev. Proteomics* **3**, 283–296.

11. Quackenbush, J. (2001) Computational analysis of microarray data. *Nat. Rev. Genet.* **2**, 418–427.

12. Hess, K.R., Zhang, W., Baggerly, K.A., Stivers, D.N., and Coombes, K.R. (2001) Microarrays: handling the deluge of data and extracting reliable information. *Trends Biotechnol.* **19**, 463–468.

13. McLachlan, G.J., Do, K.A., and Ambroise, C. (2004) *Analyzing Microarray Gene Expression Data*. Hoboken, New Jersey: Wiley Interscience.

14. Quackenbush, J. (2002) Microarray data normalization and transformation. *Nat. Genet.* **32**, 496–501.

15. Schuchhardt, J., Beule, D., Malik, A., Wolski, E., Eickhoff, H., Lehrach, H., and Herzel, H. (2000) Normalization strategies for cDNA microarrays. *Nucleic Acids Res.* **28** (article e47).

16. Yang, Y.H., Dudoit, S., Luu, P., Lin, D.M., Peng, V., Ngai, J., and Speed, T.P. (2002) Normalization for cDNA microarray data: A robust composite method addressing single and multiple slide systematic variation. *Nucleic Acids Res.* **30** (article e15).

17. Kerr, K., Martin, M., and Churchill, G. (2000) Analysis of variance for gene expression microarray data. *J. Comput. Biol.* **7**, 819–837.

18. Wolfinger, R.D., Gibson, G., Wolfinger, E.D., Bennett, L., Hamadeh, H., Bushel, P., Afshari, C., and Paules, R.S. (2001) Assessing gene significance from cDNA microarray expression data via mixed models. *J. Comput. Biol.* **8**, 625–637.

19. Cleveland, W.S. and Devlin, S.J. (1988) Locally weighted regression: An approach to regression analysis by local fitting. *J. Am. Stat. Assoc.* **83**, 596–610.

20. Kepler, T.B., Crosby, L., and Morgan, K.T. (2002) Normalization and analysis of DNA microarray data by self-consistency and local regression. *Genome Biol.* **3** (article 0037.1–0037.12).

21. Troyanskaya, O., Cantor, M., Sherlock, G., Brown, P., Hastie, T., Tibshirani, R., Botstein, D., and Altman, R.B. (2001) Missing value estimation methods for DNA microarrays. *Bioinformatics* **17**, 520–525.

22. Boutros, P.C. and Okey, A.B. (2005) Unsupervised pattern recognition: an introduction to the whys and wherefores of clustering microarray data. *Brief. Bioinform.* **6**, 331–343.
23. Eisen, M.B., Spellmann, P.T., Brown, P.O., and Botstein, D. (1998) Cluster analysis and display of genome-wide expression patterns. *Proc. Natl. Acad. Sci. U.S.A.* **95**, 14863–14868.
24. Tamayo, P., Slonim, D., Mesirov, J., Zhu, Q., Kitareewan, S., Dmitrovsky, E. Lander, E.S., and Golub, T.R. (1999) Interpreting patterns of gene expression with self-organizing maps: Methods and application to hematopoietic differentiation. *Proc. Natl. Acad. Sci. U.S.A.* **96**, 2907–2912.
25. McLachlan, G.J., Bean, R.W., and Peel, D. (2002) A mixture model-based approach to the clustering of microarray expression data. *Bioinformatics* **18**, 413–422.
26. Yeung, K.Y., Fraley, C., Murua, A., Raftery, A.E., and Ruzzo, W.L. (2001) Model-based clustering and data transformations for gene expression data. *Bioinformatics* **17**, 977–987.
27. Ng, S.K., McLachlan, G.J., Wang, K., Ben-Tovim Jones, L., and Ng, S.-W. (2006) A mixture model with random-effects components for clustering correlated gene-expression profiles. *Bioinformatics* **22**, 1745–1752.
28. Luan, Y. and Li, H. (2003) Clustering of time-course gene expression data using a mixed-effects model with B-splines. *Bioinformatics* **19**, 474–482.
29. Yeung, K.Y., Medvedovic, M., and Bumgarner, R.E. (2003) Clustering gene-expression data with repeated measurements. *Genome Biol.* **4** (article R34).
30. Guyon, L., Weston, J., Barnhill, S., and Vapnik, V. (2002) Gene selection for cancer classification using support vector machines. *Mach. Learn.* **46**, 389–422.
31. Ramaswamy, S., Tamayo, P., Rifkin, R., Mukherjee, S., Yeang, C.-H., Angelo, M., Ladd, C., Reich, M., Latulippe, E., Mesirov, J.P., Poggio, T., Gerald, W., Loda, M., Lander, E.S., and Golub, T.R. (2001) Multiclass cancer diagnosis using tumor gene expression signatures. *Proc. Natl. Acad. Sci. U.S.A.* **98**, 15149–15154.
32. Ambroise, C. and McLachlan, G.J. (2002) Selection bias in gene extraction on the basis of microarray gene-expression data. *Proc. Natl. Acad. Sci. U.S.A.* **99**, 6562–6566.
33. Bair, E. and Tibshirani, R. (2004) Semi-supervised methods to predict patient survival from gene expression data. *PloS Biol.* **2**, 511–522.
34. Beer, D.J., Kardia, S.L.R., Huang, C.-C., Giordano, T.J., Levin, A.M., Misek, D.E., Lin, L., Chen, G., Gharib, T.G., Thomas, D.G., Lizyness, M.L., Kuick, R., Hayasaka, S., Taylor, J.M.G., Iannettoni, M.D., Orringer, M.B., and Hanash, S. (2002) Gene-expression profiles predict survival of patients with lung adenocarcinoma. *Nat. Med.* **8**, 816–824.

# 13

# Biorepository Standards and Protocols for Collecting, Processing, and Storing Human Tissues

## Dean Troyer


## Summary

Recent advances in high-throughput assays for gene expression (genomics), proteins (proteomics), and metabolites (metabolomics) have engendered a parallel need for well-annotated human biological samples. Samples from both diseased and unaffected normal tissues are often required. Biorepositories consist of a specimen bank linked to a database of information. Assuring chain of custody and annotation of samples with relevant clinical information is required. The value of samples to end users is generally commensurate with the quality and extent of relevant clinical data included with the samples. Procurement of tissues is often done with parallel pre- and/or post-treatment venipuncture to obtain blood and tissue samples from the same subject. Biorepositories must also process, preserve, and distribute samples to end users. Like traditional libraries, biorepositories are meant to be used, and they are most useful when the needs of end users (researchers) are considered in the planning and development process. Ethics review and an awareness of regulatory requirements for storage, transport, and distribution are required. In the USA, Institutional Review Boards are the local regulatory entities that review protocols for banking of human biological tissues. Governmental and professional agencies and organizations provide some guidelines for standard operating procedures. The Food and Drug Administration (FDA), the Centers For Disease Control (CDC), and professional organizations such as the American Association of Tissue Banks (AATB), the American Association of Blood Banks, The International Red Cross, International Society for Biological Repositories (ISBER) and other organizations provide guidelines for biorepositories and banking of human tissues (*see* **Table 1**). To date, these guidelines are directed largely toward procurement, banking, and distribution of human tissues for therapeutic uses. In the international setting, the World Health Organization provides ethical guidelines for procurement and operating procedures.

The most commonly available tissues are formalin-fixed paraffin-embedded tissues (FFPET). While FFPET can be used for immunohistochemistry, certain DNA-based



From: *Methods in Molecular Biology, vol. 441: Tissue Proteomics: Pathways, Biomarkers, and Drug Discovery*
Edited by: B.C.-S. Liu © Humana Press, Totowa, NJ

**Table 1**
**Biobanking Resources**

| Title | Media | Publisher |
| --- | --- | --- |
| BioBank Central | Website | http://www.biobankcentral.org |
| Cell and Tissue Banking | Journal | Springer Netherlands |
| Int'l. Ethical Guidelines for Biomediocal Res Involving Human Subjects | Book and On-line | Council for Int'l Orgs of Medical Sciences, World Health Org, Geneva, Switzerlandhttp://www.cioms.ch/ |
| Collaborating with Commerical Tissue Biorepositories | Website | http://www.usm.maine.edu/ bioethics/biobank/index.html |

assays, and even RNA, frozen tissues are best suited for isolation and characterization of proteins and RNA. Freezers, back-up systems, monitors, and alarm systems and appropriate physical security are needed for long-term storage of frozen tissues. Other storage formats such as blotting onto filter paper for storage at room temperature are more commonly used for blood and can provide important "fingerprinting" for chain of custody, linking a given subject to a tissue sample.

**Key Words:** Biorepository; tissue banking; tissue preservation.

## 1. Introduction

Biorepositories have two separately managed components: The biologics bank, comprising archived frozen, fixed, and preserved tissues; and the database, including demographic data and the bank inventory. The supervisor of the physical bank is the principal investigator responsible to the Institutional Review Board (IRB) and, ultimately, to subjects. The principal investigator is responsible for the physical integrity of the samples, and must develop and supervise protocols and procedures that provide an "end to end" chain of custody from procurement through distribution to end users.

The database is independently managed, and includes modules for tracking and managing the tumor bank inventory and any new data generated by biomarker studies. The database manager is responsible for the quality, security, and integrity of the data associated with the physical biorepository. The database includes demographic, clinical, and outcome data for each specimen and tracks the bank's inventory. It may include modules for tracking additional data gathered from biomarker studies. Patient-specific data may be obtained from tumor registries, existing hospital and clinical medical records (including electronic databases) *(1)* and study records obtained at enrollment and follow-up as permitted through protocols.

Quality control of both data and samples is critical The quality of samples is dependent upon the acquisition of the correct sample, rapid sample preservation and timely transportation to the biorepository. Thus, "end to end" coordination from protocol development, consent, and sample procurement and preservation all the way through to understanding the needs of the end user are encouraged as best practices. This requires coordination with IRB personnel, clinical research nurses and study coordinators, software coders, data managers, pathologists, and surgeons on the procurement end. For distribution of samples, the needs of geneticists, basic scientists, and other end users must be understood. Banks do not exist to *store* human biologics, they exist to stimulate the discovery and validation of findings important to the detection, prognosis, and treatment of human disease. The ultimate goal is to engender improvements in health through early diagnosis, treatment, and prevention of disease. End users may include both academic/non-profit and proprietary entities.

## 2. Materials

### 2.1. Planning, Protocol Development, Ethics Review and Regulatory Environment

#### 2.1.1. Planning

The National Cancer Institute has recently published guidelines for NCI-sponsored biorepositories *(2)* that may be helpful in planning. The involvement of surgeons, clinical research associates, IRB, pathology, and researcher/end users is optimal for planning. Coordination with pathologists and surgeons involved in the project is critical, and protocols for notification of specimen availability from the operating room need to be made so that the least amount of ischemia time occurs.

#### 2.1.2. Protocol Development

A written protocol suitable for presentation to an IRB is the final product of planning and protocol development. Retention of samples for multiple studies greatly increases the value of a bank and can be ethically requested through informed consent *(3,4)*.

#### 2.1.3. Ethics Review

IRB approval will be required for both the database and the repository. New projects that utilize the data linked to banked samples are presented to the IRB for determination and may require separate IRB approval. The protocol, consent forms, and written records of communication with the IRB are filed for audits, site visits, and subsequent communication with the IRB or funding agencies. Discussions with clinical research nurses, surgeons, pathologists, database managers, and an understanding of the intellectual property guidance of the institution are also required.

### 2.1.4. Regulatory Environment

The Centers for Disease Control (CDC), Food and Drug Administration (FDA), Health and Human Services (HHS), and Occupational Safety and Health Administration (OSHA) have guidelines or regulations related to the collection and handling of human biologics (whole organs, tissues, or blood products). Of those, only the FDA has an actual inspection process with regulatory oversight for human tissue banks, and that mandate is based on human tissues and blood utilized for *transplant or therapeutic purposes.* This topic has been recently reviewed *(5).* Several states, most notably New York, have inspection and licensure processes for human tissue banks or bio-repository entities that wish to distribute tissues for transplant or research within state boundaries in addition to any state uniform anatomical gift act statutes.

Two professional associations, one based in the USA and another based in Europe, exist for the purpose of creating human tissue banking standards. The American Association of Tissue Banks (AATB) and European Association of Tissue Banks (EATB) provide standards for operation, tissue bank staff certification, and an accredited inspection process. The AATB and EATB have been very active in establishing tissue-banking standards for other countries with organizations that wish to develop human tissue banks for therapeutic, and more recently, research purposes.

## 2.2. Biorepository and Database Development and Management

### 2.2.1. Functional Roles

1. Principle Investigator: The Principle Investigator is responsible for the ethical processing, storage, and distribution of human samples and the banking activities may be for one or multiple clinical studies. These include setting policies for data management and data protection, specimen retention, and policies concerning end users and final disposition of samples. The Principle Investigator faces the dilemma of being a holder of coded information that links clinical data to specimens. The direct responsibility for procurement may rest with a separate investigator, but procurement protocols should nevertheless be planned in a coordinated fashion.

2. Database Manager: The database manager functions under the guidance of, but independently from, the Principle Investigator. One model for this function is the data safety and monitoring committees appointed to oversee the progress of clinical trials *(6)*.

## 2.2.2. Database Development

Microsoft, Oracle, and Linux are software platforms that provide scalable environments. Microsoft Access is a widely used data entry platform where tables with pull-down entries can be created. The data storage element at the "back-end" requires Microsoft SQL or a comparable system to provide reliable and secure back-up and data storage. The systems should allow for dynamic error detection notification at data entry, construction of computer forms that mimic hardcopy forms, and data lookup capabilities for subject information. A web-based data entry format allows for integration of geographically separate sites. An inventory system for the repository should be incorporated into the electronic database. A data dictionary should be part of the protocol *(7)*.

## 2.2.3. Hardware

Servers suitable for the primary computing environment are required.

## 2.2.4. Networking

External network attachments may be required.

## 2.2.5. Back-up

Back-end incorporation into a server-based system is a requirement for robust, scalable, relational systems *(8)*.

## 2.2.6. Quality Control

The first goal is simply to prevent problems before they occur. This is facilitated by the quality and training of the responsible investigators and support staff. In addition to continuous training of personnel, individuals and their analyses undergo peer review to assure both quality of science and patient care. The second goal is to detect problems by implementing routine and continuous monitoring procedures. These systems and procedures utilize data audit and report review methods to detect both random and systematic errors during the course of data collection. The third goal is to take appropriate action in a timely and effective manner to correct both the problem and its source.

### *2.3. Tissue Procurement and Preservation*

A helpful guide for materials, methods, and procedures for tissue procurement is available through the National Cancer Institute Breast and Colon Cancer Family Registries website *(9)*.

### 2.3.1. Dissection

1. forceps
2. knives
3. razor blades
4. dermatome punch biopsy 4–6 mm
5. cryomolds
6. specimen container (sterile 100 cc urine cup)

### 2.3.2. Freezing

*Cryopreservation and freezing*: Rapid immersion freezing of the sample in isopentane maintained at –20 °C (Histobath) or in liquid nitrogen is needed. It has been noted that refrigeration at 25 °C dramatically slows the degradation of RNA and may be a reasonable alternative when immersion freezing is not immediately available *(10)*. Isopentane cooled on dry ice can be used in situations where Histobath or liquid nitrogen is unavailable such as an autopsy suite or where immersion freezing is otherwise unavailable. Controlled rate freezing for specialized applications (stem cells; marrow cells; umbilical cord blood cells) can also be considered, although this is rarely utilized for freezing of tissues procured by pathologists for research uses. Controlled rate freezing is typically utilized when maintenance of viability is required *(11)*. *Preservatives*: RNA*later*™ (Ambion, Inc., Austin, TX, USA) can be used to immediately extract and stabilize tissue samples for freezing and subsequent study. A detailed protocol has been described that allows the same tissue to be preserved for gene expression and utilized for histological analysis *(12)*. Tissue samples can be linked to blood samples blotted onto paper *(13)* or buccal swabs can be used. Barcoding systems printed onto cryolabels appropriate for robust freezing thawing at –80 °C, and 2-D barcoding of aliquot tubes is available (AbGene, Rochester, NY, USA; Nunc)

Equipment for Freezing

1. Histobath for cooling isopentane to –20 °C
2. Cryostat and histology supplies for preparation of frozen section
3. Freezing container: Pyrex glass 200–500 ml beaker (size is your convenience). Note: Pyrex glass is tempered and resists breakage better than ordinary glass
4. Alternative container: Polystyrene or similar plastic beaker/container

5. Steel might work, but it absorbs a lot of heat/cool. Also, tissues adhere to it
6. Isopropyl alcohol (Isopentane). This does not have to be highest reagent grade/100%. We are not using it for chemical reactions. Some small amount of water can be tolerated. This comment is made only to anticipate cost and/or availability issues
7. Dry Ice 1–5 lbs depending upon distance traveled to procurement site, etc.
8. Styrofoam cooler to hold dry ice
9. Aluminum foil
10. Marking pen
11. Zip lock bags
12. Gloves
13. Tweezers
14. Double-edged razor blade
15. Surgical blades (these items are for handling/cutting tissue in case surgeon does not make them of appropriate size).

### 2.3.3. Transport

Transport containers suitable for dry ice. An example of validated containers for shipping frozen tissues/materials are Saf-T-PAK® (http://www.saftpak.com/home.htm). Pelleted or crushed dry ice in a styrofoam transport container is suitable for transportation. For cross country shipments by courier, validation of the amount of dry ice required to maintain temperature by doing a "dry run" may be appropriate. MVE vapor shippers (MVE Bio-Medical, Chart Industries, Inc. Marietta, GA; www.chartbiomed.com) are IATA approved. The larger shipping containers are capable of holding liquid nitrogen temperatures for up to 10 days (personal observation). These are ideal for international shipments where delays at customs or for other reasons may be anticipated.

### 2.3.4. Records

A paper record log is useful if volume permits. An electronic database is required for larger numbers and for integration with barcoding systems. We keep a small note book to record location, date, record number of subject from whom each tissue specimen is being procured.

### 2.3.5. Preservation/Fixation

RNA*later*™ (optional); 10% neutral buffered formalin; optional alcohol fixatives including Omnifix and other proprietary fixatives are available *(14)*.

### *2.4. Processing Storage and Data Entry*

### *2.4.1. Cryosections*

1. Microtome
2. Tissue Processor
3. Slides
4. OCT, Tissue TEK
5. Formaldehyde, 37–40%
6. Ethyl Alcohol
7. Hematoxylin, Biomeda
8. Eosin Y: 100 ml 5% Aq. Eosin Y, 300 ml 80% Alcohol, 2 ml Glacial Acidic Acid
9. Xylene
10. Coverslips
11. Aluminum foil; zip lock bags; marking pen.

### *2.4.2. Formalin-Fixed Paraffin-Embedded Sections*

1. Tissue cassettes (Lab-Tek)
2. Pencils for use on cassettes
3. 10% Neutral Buffered Formalin (commercial)
4. An automatic tissue processor (Lab-Tek V.I.P.)
5. Ethanol
6. Xylene
7. Paraffin (Tissue Prep and Surgipath Embedding Medium)
8. An embedding center (Lab-Tek)
9. Embedding molds (Surgipath Disposable Base Molds)
10. A precision microtome (Reichert 2030)
11. A Lab-Line or similar tissue flotation bath
12. An automatic stainer (Hacker Linear)
13. Hematoxylin (Commercial Harris formula without mercury)
14. Eosin
15. Xylene-clearing agent
16. Mounting medium (Permount)
17. No.1.5 coverglass
18. Self-Stick Printed Labels
19. Slide Containers (folders or boxes)
20. Requisition forms
21. Precleaned Microscopic glass slides (CMS Colorfrost-yellow)
22. Drying oven.

### *2.4.3. Labeling Systems*

Cryolabels appropriate for storage at −80 °C are required for sample storage in these conditions. An example is from Zebra Cryocool 300 white, Zebra material Catalog No. 5973RM.

Integrated commercial products are available that include barcoding and sample tracking methods (Freezerworks, Mountlake Terrace, WA; http://www.dwdev.com/; ItemTracker, U.K., http://www.itemtracker.co.uk/; CryoTrax freezer management and specimen-tracking system; Raleigh, North Carolina). Standard barcode readers are widely available (e.g. Symbol®; Electronic Imaging Materials, Keene New Hampshire). 2-D barcode systems are also available, and.2-D systems are particularly helpful for tray formats where up to 96 wells are printed on the bottom of tubes (ABgene; Micronics McMurrray, PA, USA) http://www.micronicna.com/).

1. –80 °C freezers with storage racks
2. Temperature monitors and alarm systems
3. Vapor phase liquid nitrogen storage system
4. Containers used for storage should be suitable for prolonged exposure to –80 °C or liquid nitrogen temperatures as appropriate. Labels must also be suitable for these temperatures and materials.

### 2.4.4. Storage, Data Entry, and Inventory Management

Pull-down "graphical user interface (GUI)" type data entry formats are helpful in reducing data entry errors. It is helpful to have the opportunity to enter limited free text as well. It is increasingly feasible to scan paper documents and store them in databases linked to the master database. Software and scanning equipment for automatic forms processing such as that provided by Teleform™ (Cardiff Software, Vista, CA, USA) can be utilized if the volume is sufficient to justify the expenditure of funds and time needed to integrate the system into a database.

## 2.5. Stability and Purity of Reagents

Chemicals labels, specification sheets, and literature provided by the supplier can be referenced for shelf life of chemicals. Reagent bottles should be marked when received and when opened.

## 2.6. Safety and Toxicity

Flammable chemicals should be stored in appropriate containers and facilities. Smaller amounts may be stored in laboratories in appropriate cabinets. A 10% neutral buffered formalin solution, commonly known as formaldehyde, is potentially carcinogenic. In 1987, precautions for laboratory workers were put in place as the Formaldehyde Standard CFR 1919.1048. Xylene is potentially hazardous and should be handled appropriately, see OSHA, 29 CFR 1910.1450. An excellent website is available for on line location of Material

Safety Data Sheets (MSDS) at Interactive Learning Paradigms Incorporated (http://www.ilpi.com/index.html). The Clinical and Laboratory Standards Institute publication GP17-A, Clinical Laboratory Safety Approved Guidelines, 2nd ed, is an important guide for laboratory safety. Guidelines for prevention of infection while handling human tissues are available from the Biosafety in Microbiological and Biomedical Laboratories published by the CDC (Appendix H) *(15)*.

### 2.7. Temperature Monitors

In many situations, the most likely scenario for failure of freezers is power interruption. In this case, the temperature monitoring and alarm systems need to function in the absence of power. In some settings, generator back-up is appropriate and required. Several types of temperature monitors are available, including those that record continuously on paper discs and recorders that electronically record, store, and alarm in case temperature parameters are exceeded.

### 2.8. Shipping, End User Agreement, and Disposal of Unused Biologics

Policies and end user agreements should be in place that assign responsibilities for handling and disposal of unused materials. Evaluation by legal departments or law firm of record is appropriate. Temperature, physical integrity, and chain of custody are key components of shipping. Shipping on dry ice should be done with validated containers shipped point to point on a test basis with temperature monitors. A variety of temperature monitors that digitally log container temperatures can provide a record of the entire transit (ElPro; Hobo).

## 3. Methods

### 3.1. Planning, Protocol Development, Ethics Review, and Regulatory Environment

#### 3.1.1. Planning

Nurses and staff of referral and pre-operative clinics are in an excellent position to determine the most appropriate time to offer participation in clinical trials. Their input is invaluable in planning the protocol and the consent form. Several patient visits are often required for evaluation and diagnosis, counseling, and pre-procedure visits. Group introduction to the study protocol through videos can be helpful for selected studies such as screening and observational trials. Coordination of record keeping and charting is also important. Notification that a surgery is scheduled, that a sample is ready, and what format end users would prefer or require as an optimal sample are all important. Banking of tissue will be much more difficult if attempted in isolation (*see* **Note 1**).

### 3.1.2. Protocol Development

The written protocol will have the following sections:

- specific aims
- background and significance
- experimental design and methods
- identification of procedures done for research vs. those as standard of care
- subject population to be studied with inclusion and exclusion criteria
- risks and measures to minimize risk
- a description of recruitment and consent procedures and any compensation
- measures to assure confidentiality
- plans for data analysis

Consent forms may be based on the model form of a cooperative multicenter trial or follow local IRB guidelines.

### 3.1.3. Ethics Review

Ethical issues specific to the academic and proprietary uses of human biological samples have been recently reviewed *(16)*. Consent and protocols must follow IRB guidelines and typically include a request for retaining tissue for research uses and indicating that both for-profit and non-profit entities may utilize the tissues. A frequent IRB requirement is that subjects be allowed to withdraw samples from the biorepository if they choose. It may be appropriate to discuss obtaining separate IRB approval for the physical repository and for the database with IRB officials (*see* **Note 2**).

### 3.1.4. Regulatory Environment

The International Committee on Harmonization (ICH) Good Clinical Practice guidelines E6(R1), FDA Regulations relating to good clinical practice and clinical trials (CFR 21), and FDA Good Laboratory Practice guidelines may be helpful references. New projects that use an established biorepository are submitted for determination of review status (e.g., exempt, expedited, full board review).

### 3.2. Database Development

There are two components of a biorepository, the physical repository for biologics and the associated database (*see* **Note 3**). The physical repository consists of frozen and fixed tissues and other biologics that have been accessioned into the bank.

### 3.2.1. Functional Roles

1. Principle Investigator: The management and policies of the Utah-based Resource for Genetic and Epidemiological Research were recently described and compared to those of deCode Genetics *(17)*. In brief, there is a balance to be sought between data security, linking data to coded samples, and accessibility. Coding of samples requires a key and a keyholder, and it has been proposed that a data protection officer function be created to deal with the assignment and entry of codes *(18)*. At the UTHSCSA Tissue Bank and Data Base for Urologic Diseases, we have created a function similar to this whereby the Principal Investigator does not have access to coded information, a role carried out jointly by an archivist who is also an experienced histotechnologist, and the database manager.

2. Database Manager: Best practices for database development and management include separately delineated responsibilities for system developers (coders), data managers, and the Principle Investigator of the biorepository *(8,19–21)*. The supervisor of the biorepository is responsible for the physical integrity of the samples and assuring the quality and integrity of the database and an independent data manager is responsible for the integrity of the associated data (*see* **Note 4**). For large databases, the functions of data management have been further subdivided. Institutional academic data management groups may be an excellent source of these coding, management, and compliance functions. These responsibilities should be identified and assigned as part of the planning for the bank and biorepository. Conflicts and disagreements are difficult to resolve after the fact in the absence of a structure and policies (*see* **Fig. 1**).

### 3.2.2. Database Development

An electronic database provides the operational support for most biorepositories. The database has two components: an inventory system and clinical and outcomes data elements utilized in projects that are linked to tissue and other biologics in the physical repository. Treating these separately has advantages, including the ability to enter and record changes in the inventory without perturbing the clinical data element portion of the database. The database includes demographic, clinical, and outcome data for each specimen and will keep track of the bank's inventory. Patient-specific data can be obtained from existing data sources, primarily previous study records accrued at enrollment and follow-up, hospital tumor registries, and hospital and clinical medical records (*see* **Note 5**). The database should also include modules for tracking and managing the tumor bank inventory and any new data generated by biomarker studies.

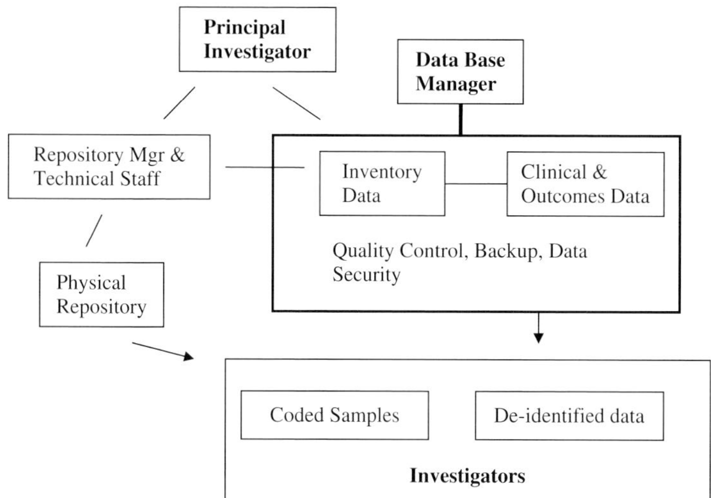


Fig. 1. An outline that shows the relationships and functions of an independent database manger to the principal investigator.

A simple log book can be used to record specimens. However, the needs of a biorepository, especially when the number of samples increases to many thousands, can quickly overwhelm manual tracking methods. Barcoding systems can be helpful and can be readily incorporated into most databases.

Databases must track collection, processing, and distribution of specimens and must be searchable *(8)*. Scalable relational databases with the capacity to incorporate disparate types of data are required *(8)*. Linking results from biomarker studies back to the existing samples requires flexibility to incorporate disparate types of high-throughput data from expression arrays, metabolomics, genomic and epigenomic assays, mass spectroscopy, and proteomics. Both desk-top and server-based systems are available. Microsoft Access is a widely used desk-top database. Server-based systems include Microsoft SQL and Oracle. The operating database must also provide reports and be accessible to appropriate users. Graphic user interfaces (GUI) with pull-down menus are helpful for both data entry and utilization by scientists (*see* **Note 6**).

Providing access to the data linked to specimens in the biorepository while assuring privacy and security is challenging. A committee comprising the biorepository supervisor, clinical colleagues who have contributed to the repository, and others knowledgeable about the applications of data to research is suggested *(22)*. The Health Insurance Portability and Accountability Act (HIPPA) must also be heeded in handling and distributing data *(23)*.

### 3.2.3. Hardware, Networking, Back-up, Quality Control

The final product should be well-documented and address hardware, networking capabilities, network and database security, and quality control. Protocols for daily and weekly back-ups are required. Daily back-ups onto tape or CDs should be stored in fire-resistant vaults on-site. Weekly off-site storage of back-up media is part of best practice.

A web-based approach to data access can be considered and issues such as access control, security, and levels of access must be addressed *(24)*. Scalability and back-up capabilities are among the issues that may influence decisions at this level. Further guidance from the Society of Clinical Research Associates and published literature *(25)* can assist with system documentation, system and database security, maintenance of data integrity, regulatory, and compliance issues *(23)*.

## 3.3. Procurement, Preservation, Storage, Shipping

The activities of the bank should be conducted following a set of standard operating procedures such as those provided by the International Society of Biological and Environmental Repositories (ISBER) *(26)* (*see* **Note 7**). In addition to the ISBER, the American Association of Tissue Banks (AATB) provides guidance for standard operating procedures in its Self-assessment Tool/Audit Report (STAR) for therapeutic tissue and reproductive banks, Section B, General Organizational Requirements of a Tissue Bank *(27)*.

Data generated from translational studies are affected by both pre-analytical and analytical variables. Pre-analytical variables are those related to sample procurement, processing, transportation, or storage that affect the analytical result. Analytical variables are those directly involved in the assay itself such as instrumentation and standardization. Many tissue samples are obtained by surgical excision, and during surgery, vessels are clamped or extirpated to control bleeding. Operative time varies for different types of samples. The total ischemia time is the time between devascularization and freezing or preservation of tissue. Ischemia time can affect gene expression and protein levels *(28)*. Samples such as biopsy cores can be obtained and frozen much more rapidly than samples from a resected organ. Skin and muscle biopsies can be frozen within seconds of procurement. By contrast, a breast biopsy might require 20–30 min operative time and the intraoperative time for radical retropubic prostatectomy is approximately 2 h *(29)*. Thus, in some cases, more ischemia time will be required than others. Surgeons alert the operating room circulator that samples for biobanking are to be delivered to the pathology department in the fresh state rather than in fixative (*see* **Note 8**).

### 3.3.1. Dissection

The specimen is weighed, measured, and photographed according to protocols in place for the diagnostic histopathology laboratory (*see* **Note 9**). By gross examination, areas of normal and abnormal tissue are sampled using a 6- to 8-mm dermatome core biopsy, a procedure that allows localization of the sample by virtue of the resulting defect. Other options include using a biopsy gun (18 G) to obtain small cores of tissue that do not disrupt architectural relationships. If circumstances permit, a frozen section may be performed on the targeted disease tissue to confirm amounts of diseased tissue present and accuracy of sampling. Tissue-sampling locations and amounts must not interfere with routine diagnostic and staging (*see* **Note 10**). These cores are frozen immediately in the isopentane bath or liquid nitrogen Dewar flask (*see* **Note 11**). The frozen tissues are transferred to a container, placed on dry ice, and transported immediately back to the laboratory. (*see* **Note 12**).

### 3.3.2. Freezing

Tissue sections/cores are immersed in isopentane cooled to –20 °C or liquid nitrogen (*see* **Note 12**). Thickness of tissues should be 5–10 mm thick by 1.5 × 1.5 cm to allow for quick freezing. Tissues may remain in isopentane for long periods without harm. A minimum of 2–3 min should be allowed for complete freezing. Specimens are then transferred to a container with dry ice for transport to the biorepository.

If a table top histobath or liquid nitrogen are not available, the following procedure can be used:

1. Place freezing beaker/container in dry ice. Dispense 100–250 ml isopentane (sufficient to cover tissue to be frozen)
2. Allow isopentane to cool down while sitting on dry ice for at least 15 min. (If traveling to site with isopentane, a screw cap bottle or whatever stores the isopentane, if small enough, can be placed on the dry ice before traveling to site. This would assure cooling in advance.)
3. If portions of tissue are greater than 1 cm in maximum thickness, slice them to this thickness to allow more rapid cooling. Tissue length/width simply needs to allow immersion in whatever volume of isopentane present in freezing beaker. Think of a 5- to 8-mm thick postage stamp as an optimal size of tissue to freeze.
4. Five minutes immersion is sufficient to freeze tissue. Check for firmness by grasping with gloved fingers or probing with tweezers. If it feels solid, it is frozen.
5. Prepare squares of aluminum foil 5–10 cm length/width. Using a permanent marking pen (Magic Marker is fine), label these squares with a number or something that identifies the sample with the subject. You may also need to identify the location from where the particular sample is removed if receiving multiple tissue samples from the same subject. Place the label on the foil square

so that when the foil is folded over the tissue, the label is visible on the surface. Tightness of the foils and folds is not critical: just fold it to make sure the tissue stays enclosed.

6. Place the frozen tissue on the foil. Fold the aluminum foil over the tissue. Foil-wrapped tissues can be placed directly on dry ice or placed in a zip lock, which can then be placed on dry ice for transport back to laboratory.

Place the tissue in a zip loc bag that is also labeled with the subjects ID number/barcode as needed. If freezing multiple pieces of tissue from same subject, foil wrap each piece individually and place these in the one zip lock plastic bag if they will fit. If they would not fit, you may need to use multiple zip lock containers. This technique allows several locations of the tumor or the tumor plus a lymph node from the tumor (etc.) to be frozen. The location of the tumor should be recorded on the foil if there are multiple samples. Basically, you can write anything on the foil that is needed as long as it fits.

### 3.3.3. Transport

Specimens kept on dry ice should be transported as soon as feasible to the biorepository (*see* **Note 13**).

### 3.3.4. Records

Clinical facility location, date, record number of subject from whom specimen is being procured, and other information are recorded. Additional demographic and diagnostic information may be subsequently obtained from the medical record or the laboratory information system after the diagnostic pathology report is completed.

### 3.3.5. Preservation/Fixation

Commercially available preservatives or fixatives such RNA*later*™ (Ambion), and alcohol-based fixatives such as Omnifix are available. Ten percent neutral buffered formalin is the standard widely used fixative for preserving morphology. Samples should be fixed in 10 vol or more of a formalin- or alcohol-based fixative. For the best fixation, samples should not exceed 1–2 mm in thickness.

Dissect the tissue sample selected for the biorepository as rapidly as possible. Place in 10 vol of fixative at room temperature. After 1 h, inspect the sample. If the fixative is cloudy, replace it with fresh fixative. If the sample is thick, remove sample, place on a cutting board, cut 1- to 2-mm slices to expose all surfaces and return to 10 vol of fixative. (Note: a short period of fixation will firm up hard-to-slice tissues such as lymph nodes, spleen, and thymus, making them easier to slice). Routine fixation should be for 6–12 h of over night fixation. Following fixation, the specimen should be removed and placed in 70% ethanol until shipping or further processing into paraffin.

An alternative to freezing or preservation in liquid fixation/preservation is imprinting tissue on nitrocellulose paper *(30,31)*. Cross-sections of fresh tissue are directly imprinted onto nitrocellulose paper. This imprints cells onto the paper from the tissue and also preserves localization within the cross-section. This technique provides a rapid and inexpensive means to preserve dried cells without further fixation or refrigeration. Both DNA and RNA can be recovered from these imprints.

Immediate stabilization and preservation of RNA is a prerequisite for accurate gene expression analysis. Alterations in gene expression triggered by sample manipulation must be prevented. Immediate protection in tissue with RNA*later*™ is essential. RNA*later*™ [Ambion, Catalog Nos. 7020 (100 mL), 7024 (250 mL), 7021 (500 mL)] is an aqueous, non-toxic tissue storage reagent that rapidly permeates tissues to stabilize and protect cellular RNA. RNA*later*™ eliminates the need to immediately process tissue samples or to freeze samples in liquid nitrogen for later processing. Tissue pieces can be harvested and submerged in RNA*later*™ for storage without jeopardizing the quality or quantity of RNA obtained after subsequent RNA isolation.

1. Aliquot 2 ml of RNA*later*™ (Ambion, Applied Biosystems Foster City, CA) into a 4-ml cryogenic vial (Corning, Catalog No. 430662, Lowell, MA). (If tissue weight is to be determined, it will be necessary to weigh the collection vial with RNA*later*™ before introducing the tissue sample into the collection vial. It is essential that the tissue be preserved in RNA*later*™ immediately. The tissue weight can be derived by weighing the collection container with RNA*later*™ and tissue and subtracting the weight of the container with RNA*later*™ without tissue.)

2. Excise tissue sample. Trim tissue as necessary so that at least one dimension measures less than 0.5 cm (5 mm or ~0.2 in.). The tissue sample should be no larger than 0.4 g (400 mg).

3. Submerge the tissue sample in the RNA*later*™ and label the vial with the appropriate barcode label.

4. As soon as possible (within 1 h), transfer the cryogenic vial containing the tissue sample in RNA*later*™ to a 4 °C refrigerator. Leave the sample at 4 °C from 4 h to overnight, to allow the RNA*later*™ to penetrate the tissue.

5. Following the incubation at 4 °C, transfer the sample vial to a –70 °C freezer until ready to ship to the Sample Collection and Coordination Site. (Note: If transfer of the sample to a –70 °C freezer the next day is not practical (e.g., for tissues harvested on a Friday), it is acceptable to leave the sample at 4 °C for up to several days.)

### *3.4. Processing*

#### *3.4.1. Cryosections*

Properly orient and embed specimen in OCT (*see* **Note 14**). Quick freeze by submersing in isopentane cooled to –55 °C in the histobath. Mount specimen into the cryostat. Cut 5-µm thick sections on plus coated slides, labeled with the patients name or accession number. Fix in 10% formalin in 80% alcohol for 1 min. Rinse 10 dips in deionized water. Stain with Biomeda hematoxylin for 30 s. Rinse with deionized water. Counterstain with Eosin Y for 30 s. Dehydrate, clear in xylene, and coverslip.

#### *3.4.2. Formalin-Fixed Paraffin-Embedded Sections*

The specimen is received into the laboratory along with a completed requisition indicating: the specimen identification, number of cassettes, researcher name, department, account, telephone extension, procedures to be performed on the specimen, and any additional pertinent information. The requisition is logged into the proper log book (research or patient) and given a laboratory number for identification. Then the cassettes are placed on the automatic processor (overnight or day run) to be dehydrated, cleared, and infiltrated with paraffin (*see* **Table 2**). When the processor has finished, the cassettes are removed and placed in a paraffin bath to await embedding. The specimen is

**Table 2**
**Representative Processing Sequence for Formalin-Fixed Paraffin-Embedded Tissue Using an Automatic Tissue Processor**

| Automatic Tissue Processor | Processing Schedule |
| --- | --- |
| 1. 70% Reagent Alcohol | 30 min |
| 2. 80% Reagent Alcohol | 30 min |
| 3. 95% Reagent Alcohol | 30 min |
| 4. 100% Reagent Alcohol | 45 min |
| 5. 100% Reagent Alcohol | 45 min |
| 6. 100% Reagent Alcohol | 45 min |
| 7. Xylene | 45 min |
| 8. Xylene | 45 min |
| 9. Xylene | 45 min |
| 10. Paraffin 60 °C | 45 min |
| 11. Paraffin 60 °C | 30 min |
| 12. Paraffin 60 °C | 30 min |
| 13. Paraffin 60 °C | 60 min |

removed from the cassette, and a mold is selected according to the size of the tissue. The mold is filled with molten paraffin (Polyfin TBS, Catalog No. Y10842), then the tissue is placed into the mold, the cassette bottom placed on top, filled with paraffin, and allowed to solidify on the chilled portion of the embedding center. The resultant paraffin block is sectioned with a precision microtome and the sections placed on clean glass slides. The sections are then dried in an oven at 60 °C for 20 min (minimum) before hand staining or placed in the automatic stainer that has an area for drying the slides before staining. The resultant stained slide is coverslipped, labeled properly, and placed in a slide container for delivery to the requesting party. The requisition is completed and filed appropriately.

### 3.4.3. 3.4.3. Labeling Systems

Advance preparation of specimen kits that include aluminum foil, storage bags, and barcoded labels may be helpful (*see* **Note 15**).

### 3.4.4. Storage Data Entry and Inventory Management

Portions of tissue including normal and tumor-containing tissue are wrapped in aluminum foil, labeled with a permanent marker (Block 1,2,3, etc.), placed into a small zip lock bag and placed into a small sliding lid box. These are then stacked inside of larger cryoboxes that are labeled (*see* **Note 16**). The timing for data entry is best determined by work-flow for given sites. If the system is web-based, data entry can occur at locations convenient to after hours or off-site situations.

### 3.5. Stability and Purity of Reagents

The longevity of analytes in frozen samples is not well-known, for fairly obvious reasons. However, freeze–thaw cycles are known to induce changes in sample stability, and work should be planned to avoid thawing of tissue except as it is being prepared for immediate extraction of protein, DNA, or RNA. Histological and other reagents are shipped with specifications that include shelf life.

### 3.6. Safety and Toxicity

Institutional training programs for handling toxic wastes should be utilized. On line resources and federal regulations are referenced in **Subheading 2.7**.

### 3.7. Temperature Monitors

Monitors should be in place that record and alarm with notification of a call list. The most common problems are brief power outages, condenser coils clogged by dust, and opening and closing doors that trigger alarms.

### 3.8. Shipping, End User Agreement, Disposal of Unused Biologics

#### 3.8.1. Requests to Access-Banked Tissues

Access to banked material is strictly controlled and will require approval by the bank's executive committee. The latter includes members of the collaborating groups and the Bank Director. Any requests for patient identifiers require separate IRB approval.

A written form requesting use of tissue from the biorepository should be available. The written request must include biologics requested, preservation methods, donor criteria, and evidence of institutional approval for the proposed use. A committee that operates from written procedures should review requests. An end-user agreement is used for distribution of approved tissues to end users. A Master Materials Transfer Agreement housed at the Association of University Technology Transfer Managers is a helpful reference document for this purpose *(32)* (*see* **Note 17**).

#### 3.8.2. Shipping

Although formalin-based fixatives are accepted, they cannot be shipped through the mail without a permit. Samples initially placed in a non-alcoholic fixative should be dehydrated in alcohol and shipped in either 70% alcohol or a suitable buffer. Alcohol-fixed samples are not considered biohazards and may be sent through the US mail without special processing or labeling. Standard shipping of frozen tissues and extracts is generally done on dry ice (*see* **Note 18**).

Nitrogen Vapor Shippers are available in a variety of sizes. They are designed for safe transportation of biological samples around the world and are IATA approved for this purpose. A hydrophobic absorbent holds the liquid nitrogen and allows storage and shipping at cryogenic temperatures ($\leq 150\,°C$) for 10 days or more without recharging. (We have shipped serum in an MVE shipper that maintained a temperature of $-180\,°C$ for 10 days—author note.) The units can be monitored and recharged in the same fashion as liquid nitrogen storage systems. A protective shipping container is available that protects the container during transport and prevents it from being placed on its side. These are available through a variety of distributors including Praxair and MVE Bio-Medical, Chart Industries, Inc. (*see* **Note 19**).

### 3.8.2. Disposal of Unused Product

Policies and the end user agreement should specify the disposition of unused product once a research project has been completed. Destruction by incineration is the most robust method and should be considered as a standard. Commercial firms or institutional facilities may be available for this purpose. Institutions usually have agreements with commercial providers to handle and dispose of human and other biological materials through incineration. Biorepositories engaged in banking of human cells, tissues, and cellular and tissue-based products for therapeutic applications in humans are required to file form 3356 under 21 CFR Parts 1271, 207.20, and 807.20. Research Biorepositories do not fall under these regulations; however, these exemplify the highest standards for banking of tissues.

## 4. Notes

1. Nobody wants to pay for biorepositories, but everyone wants to use them. Rarely do institutions see sufficient value in developing a biorepository that money will flow in support. Therefore, bartering and persuasion are the order of the day in many cases when efforts first begin. This involves relationships amongst pathologists, surgeons, and end users (researchers). Forming a team whose mission is to build the biorepository is critical. Everyone needs to contribute, and everyone involved needs to see some form of benefit. Because of daily diagnostic obligations, pathologists may be overwhelmed and reluctant to procure or sacrifice tissue for purposes that seem remote. Hospitals and health care systems operate inside boxes surrounded by walls and are reluctant to do anything that will have a cost. This is understandable, and so, success proceeds out of diplomacy and persistence.

2. Ethics review processes are a source of puzzlement and even frustration. If translational research is to move forward, those building biorepositories need to make the barriers for use of the samples as low as possible. Only limited review by the IRB may be required if the tissues in the repository are already under an IRB protocol, and the end user does not require and will not receive identifying information. Consult your IRB Chair and administrative personnel. Make friends of them. Institutional IRBs are generally responsive to the needs of the research community and capable of distinguishing between therapeutic interventional protocols and those involving use of human biologics with no identifiers.

3. Database design is easily afflicted by the pursuit of perfection. If too many data elements are required, the effort required to enter specimens into the bank quickly exceeds the available time. If resources are unlimited, which they seldom are here on planet earth, then data could be pursued in an unlimited way. Balance practicality against idealism.

4. A data manager independent of the interests of the principal investigator and clinical project manager should be the custodian or steward of the database. The role is exemplified by the function of a statistician/data manager who sits on a data safety and monitoring committee for a clinical trial. Even when resources are limited, the integrity of the database should be carefully guarded. Just as in clinical trials, there should be an audit trail/time date stamp for changes that are made in the database.

5. The format of the existing clinical data to be linked with the specimens is a significant issue. If the data are in a transferable electronic format, the effort to procure that data is less than it is for paper records. When resources are limited, it may not be possible to collect all the information desired.

6. The working database accessible on a daily basis for new entries is periodically rolled into the database of record that is maintained with quality control measures by the database manager. Regular back-ups are mandatory. The worst possible scenario is loss of all data because of catastrophic events, so off-site back ups of media containing the data is mandatory. Maintenance of data integrity through quality control, back-ups, and fire walls requires expertise and resources—or committed colleagues with the relevant expertise. Databases can be the source of angst as issues of control and ownership come to the fore. Therefore, think carefully about the architecture of responsibility before launching the repository. If institutional resources for computing and database management are limited, you may need to be creative. If institutional resources exist, cultivate friendships. Do not carry the database on a laptop or thumb drive—witness the number of cases where stolen laptops have compromised confidence in the ability to maintain database security. Specimens and paper records should be stored in locked rooms that are secure against intrusion.

7. A policy governing how samples discovered to be derived from subjects with infectious diseases should be developed. Because most tissue banks do not routinely screen donors for HIV and hepatitis, it is often after procurement that such tissues are recognized as potentially infectious. Universal precautions for handling tissue should be in place. In other words, human tissues should always be handled as if they are potentially infectious. However, it is interesting to note that therapeutic tissue banks extensively test subjects for at least hepatitis and HIV during the procurement process. Only rarely is this considered prospectively in research banking, usually because of limited resources. End users must be aware that tissues could be infectious, and policies should be in place to deal with notification and disposition of samples once it is known that a donor has an infectious disease such as hepatitis or HIV. Some banks of course are explicitly developed to house biologics from such patients. It is simply a matter of communication and policy.

8. For human biologics, a pathologist is almost indispensable to procurement. Pathologists receive all tissues removed from patients in the operating room and therefore are in the pipeline for receiving tissue samples. Getting the samples into the hands of an interested pathologist in turn requires an interested and motivated

surgeon. It is worth noting that the surgeon should welcome the pathologist's role as steward of the tissue. This isolates the interests of the patient and their surgical treatment from any interests that arise from sample procurement. The potential for conflict of interest is therefore minimized when someone other than the surgeon selects which and how much tissue is procured for banking. The operating nurses and histotechnologists can make or break procurement of tissue. Once these folks are your friends, they will notify a relevant pathologist as soon as the tissue is available in the operating room.

9. Sterility and cross contamination during dissection are a concern. Disposable absorbent pads and even paper towels can be laid out to create a field that is at least clean if not sterile. Sterile blades and sterile gloves come in packaging that can be opened and laid out to create a small cutting surface that is sterile. The use of disposable knives will prevent cross contamination between samples. Beakers of bleach and 95% ethanol can be used to "sterilize" forceps and other non-disposable dissection tools between samples.

   Tissue samples (diseased/non-diseased; tumor/non-tumor) should be approximately $1.5 \times 1.5 \times 0.5$ cm. For freezing, OCT compound is used to embed the frozen tissues. (There may be special applications where OCT is not desired, and this should be considered in advance.) For specimen dimensions, think of a thick postage stamp. The flat surfaces readily lend themselves to cryosectioning. These will freeze quickly and can be readily sectioned in a standard cryostat.

10. Cores of tissue (5–8 mm in diameter) obtained with a disposable dermatome biopsy can be frozen intact lying horizontally or "on end." Note, however, that if the cores are lying horizontally, it will be necessary to section away and waste significant amounts of tissue before obtaining a full section of the core. Alternatively, if the tumor or diseased tissue is small, the core can be divided longitudinally and some frozen, some submitted for paraffin embedding and routine histology, and some harvested directly for RNA/DNA isolation.

11. Biorepository samples are often obtained from a larger resection sample. It can be very challenging to maintain the mapping and orientation of the biorepository sample to its "parent" resection. This seems simple, but it is difficult to enter paper data or handwritten notes into a permanent electronic database, and paper descriptors/maps get cumbersome very quickly. A small paper strip can be embedded in one corner of the tissue cryomold to serve both as an identifier and as a means to orient the tissue. Paper diagrams or "maps" can be used to locate the tissue within a larger specimen. Permanent inks can be used to designate "anterior/posterior/lateral/medial" or similar 3-D orientation. Photographs can be used to document the larger sample from which the biorepository sample is taken. Digitized photos can be easily stored. Mammography or X-ray machines are sometimes available in pathology suites. An X-ray of the tissue specimen can also be used to annotate the location from which biorepository samples are obtained. The prostate poses one of the more challenging situations for "mapping" because there are multiple primary tumors with separate lineages in most prostates removed for primary treatment of prostate cancer. In contrast,

most colo-rectal, breast, and lung cancers do not pose the same challenge because multiple primary tumors are not typically present in the resection specimen.

12. Personnel asked to participate in banking studies are often intimidated by the lack of equipment and material for freezing. The first thing often noted as lacking is liquid nitrogen. Freezing in liquid nitrogen is not necessarily the optimal approach. Isopentane cooled to $-20\,°C$ in a bench top Histobath or in a freezer is a suitable medium for rapidly freezing tissues. In fact, it introduces fewer bubble artifacts into the tissue than does liquid nitrogen. A beaker of isopentane sitting on dry ice can be used in settings where refrigeration is unavailable. Notes and Precautions:

- Maintain samples on dry ice for transport back to the laboratory.
- Remember that cryostats (sometimes present in laboratories where tissue is processed) and standard freezers frequently have daily/weekly defrost cycles built in. It is not safe to leave samples in these overnight.
- Isopentane is an alcohol and flammable. It is not as volatile nor as flammable as benzene or ether. However, when storing it in a freezer for long term, the container should be closed or covered with foil.
- Appropriately labeled zip lock bags will be the final storage mechanism. They can be lined up in small boxes to keep them upright and in order.
- Disposable tissue molds serve well to receive tissue sections or cores for freezing. These are typically disposable plastic. They are wrapped in aluminum foil for transport or shipping because they can fall out of their cryomolds.

13. A styrofoam box can be used to transport the specimen. Pelleted dry ice that can be heaped up around the specimen container is ideal. Alternatively, broken up dry ice as opposed to large chucks or blocks is preferable. Urine cups with screw on caps make ideal sample collection transport containers. Once tissues are frozen, the cup can be labeled, specimens placed inside, and the lid screwed on securely. The container can then be placed on dry ice for transport to the laboratory.

14. There are biorepositories that blindly bank tissues without preparing histological or cryosections. We recommend that all tissues entered into a biorepository, frozen or otherwise, be sectioned for microscopic examination. The presence and amount of tumor can be assessed in each sample. Do not assume that just because something is labeled tumor that it is tumor/diseased. A small tip permanent marker can be used to circle areas on the slide for laser capture, scraping, and also, a date and study code can be recorded to indicate when material was utilized from the corresponding block.

15. Labeling and entry of samples into the repository is where errors can occur. It can be helpful to prepare kits with bags, cassettes, and labels in advance for use by the procurement pathologists. Tubes, containers, bags, and even paper can be barcoded for subsequent use and scanning for entry and linkage to a database.

16. These boxes then slide into a tower or rack suitable for storage in the –80 °C freezer or in a liquid nitrogen system. The vapor phase storage systems are very attractive alternatives to liquid nitrogen both for shipping and storage. Although liquid nitrogen is well-entrenched, it is worth carefully examining vapor phase nitrogen storage systems. The labeling systems can be very sophisticated using barcodes or simple. We think of the sample location as analogous to a room in a hotel. Each hotel (tower or rack) has multiple floors, and there are multiple rooms on each floor. Tissue samples from each subject, even if multiple, are stored in one "room."

17. The database should have a means to record distribution of tissue to an end user. This could be another column in a ledger, or it could be an area where text can be entered into an electronic database. We find that cryosections themselves serve as a record of what is in the tissue, what tissue was divided for use by a researcher, and what tissue remains for the next use. Permanent marker can be used to circle tissue on the glass slide of the stained, coverslipped cryosection removed for research along with the date.

18. We prepare a frozen (cryosection) before and after preparation of a sample for distribution to a researcher. Working in a cryostat, a single edge razor blade works nicely to cut the tissue embedded in the OCT. The portion of tissue cut for the investigator is wrapped in aluminum foil and placed in a small labeled zip lock bag.

19. International Shipping:

    - If biologics are not known to be contaminated (e.g., screened and negative for HIV 1 & 2, hepatitis B & C, Syphilis, gonorrhea, and TB) and no positive samples would be sent to the USA for research purposes. The FDA has no jurisdiction over operations.

    - Importation into the USA is governed by the USDA, which delegates human biologics to the CDC, import permit division. An application for an import permit is made to this division. Either a permit or a memorandum indicating that a permit is not needed is issued. These documents are attached to each shipment and/or filed with the Port Director at the anticipated port of entry.

    - Recent Patriot Act legislation and associated regulations are relevant to importation from international sites. The TSA is tasked with security of shipments, and all their internal regulations concerning known shipper status are classified as security information. TSA only oversees carriers, not known shippers. Although there are approximately 1 million "known shippers" in the US, TSA monitors the 4300 carriers. 4000 of these are intermediaries such as QuickStat that use passenger planes and the other 300 such as Air Net are direct cargo only carriers. TSA's job is to prevent dangerous materials from getting on planes. Once on a plane, they have no further hand in the shipment. Customs handles things on the receiving end. If TSA has an issue with a particular shipment, they first contact the carrier and ask to see the "known shipper" documentation in their records. It is incumbent on the carrier to have the known shipper documentation on hand.

- Once the material is in the USA, the same security issues apply if we ship the material by air to anyone else whether that is domestically or exported.
- Site visits to become "known shipper" should be documented. Entities may think they are known shippers and in fact the company that did the site inspection never recorded the information internally.
- Document and record all requests for information and data.
- Contact the Assistant Director for Inspection Oversight, TSA Office of Compliance.
- Customs is the responsibility of the carrier used to bring shipments into the USA. If CDC and TSA requirements are met and all the necessary documentation is on hand, this is all that will be required. The carrier will handle getting shipments through customs.

## References


1. Goodman, K.W. (1996) Ethics, genomics and information retrieval. *Comput. Biol. Med.* 26: 223–229.
2. http://biospecimens.cancer.gov/biorepositories/guidelines_full_formatted.asp
3. Wendler, D. (2006) One-time general consent for research on biological samples. *Arch. Int. Med.* 166: 1449–1452.
4. Hansson, M.G., Dillner, J., Bartram, C.R., Carlson, J.A., and Helgesson, G. (2006) Should donors be allowed to give broad consent to future biobank research? *Lancet Oncol.* 7: 266–269.
5. AORN Recommended Practices Committee (2006) Recommended practices for surgical tissue banking. *AORN J.* 83: 435–442; 445–450; 453–456.
6. Ellenberg, S.S. and George, S.L. (2004) Should statisticians reporting to data monitoring committees be independent of the trial sponsor and leadership? *Stat. Med.* 23: 1503–1505.
7. van Es, G.A. (1996) Research practice and data management. *Neth. J. Med.* 48: 38–44.
8. Bao, W., Schmid, J.E., Goetz, A.K., Ren, H., and Dix, D.J. (2005) A database for tracking toxicogenomic samples and procedures. *Reprod. Toxicol.* 19: 411–419.
9. http://epi.grants.cancer.gov/CFR/biospecimen_fresh_tissue.html#safety
10. Almeida, A., Thiery, J.P., Magdelenat, H., and Radvanyi, F. (2004) Gene experession analysis by real-time reverse transcription polymerase chain reaction: influence of tissue handling. *Anal. Biochem.* 328: 101–108.
11. Ketheesan, N., Whiteman, C., Malczewski, A.B., Hirst, R.G., and La Brooy J.T. (2004) Effect of cyropreservation on the immunogenicity of umbilical cord blood cells. *Trans. Apheresis Sci.* 30: 47–54.
12. Florell, S.R., Coffin, C.M., Holden, J.A., Zimmermann, J.W., Gerwels, J.W., Summers, B.K., Jones, D.A., and Leachman, S.A. (2001) Preservation of RNA for functional genomic studies: A multidisciplinary tumor bank protocol. *Mod. Pathol.* 14: 116–128.

13. Davis, J.C., Shon, J., Wong, D.T., Jaffe, S., and McEvoy, J.A. (2005) DNA-based biological sample tracking method. *Cell Cell Presev. Technol.* 3: 54–60.
14. Titford, M.E. and Horenstein, M.G. (2005) Histomorphologic assessment of formalin substitute fixatives for diagnostic surgical pathology. *Arch. Path. Lab. Med.* 129: 502–506.
15. Biosafety in Microbiological and Biomedical Laboratories, U.S. Department of Health and Human Services, Centers for Disease Control and Prevention, and National Institutes of Health, 4th ed., May 1999, available online at http://www.cdc.gov/OD/OHS/biosfty/bmbl4/bmbl4toc.htm.
16. Nestler, G., Steinert, R., Lippert, H., and Reymond, M.A. (2004) Using human samples in proteiomics-based drug dev elopement: bioethical aspects. *Exp. Rev. Proteomics* 1: 77–86.
17. Wylie, J.E. and Mineau, G.P. (2003) Biomedical databases: Protecting privacy and promoting research. *Trends Biotechnol.* 21: 113–116.
18. Auray-Blais, C. and Patenaude, J. (2006) A biobank management model applicable to biomedical research. *BMC Med. Ethics* **7**:4.
19. Eiseman, E., Bloom, G., Brower, J., Clancy, N., and Olmsted, S.S. (2003) Case Studies of Existing Human Tissue Biorepositories, The Rand Corporation. Santa Monica, CA, www.rand.org/pubs/authors/e/eiseman_elisa.html.
20. Gregersen, P.K. and Lundsten, R. (2001) Robotic and Information Systems for large scale Genetic Studies. The North Shore LIJ Biorepository and the New York Cancer Project. Biotechnic and Histochemistry 76: 223–232, see also http://www.biorep.org/Library.
21. Burnett, L., Barlow-Stewart, K., Proos, A.L., and Aizenberg, H. (2003) The "Gene Trustee": A universal identification system that ensures privacy and confidentiality for human genetic databases. *J. Law Med.* 10: 506–513.
22. Wylie, J.E. and Mineau, G.P. (2003) Biomedical databases: Protecting privacy and promoting research. *Trends Biotechnol.* 21: 113–116.
23. Gunn, P.P., Fremont, A.M., Bottrell, M., Shugarman, L.R., Galegher, J., and Bikson, T. (2004) The health insurance portability and accountability act privacy rule – A practical guide for researchers. *Med. Care* 42: 321–327.
24. Blobel, B., Nordberg, R., Davis, J.M., and Pharow, P. (2006) Modelling privilege management and access control. *Int. J. Med. Informatics* 75: 597–623.
25. Connell, C.A. (2003) ERES compliance and data integrity issues in e-Trials. SoCRASource, February:24–27, available online at http://www.socra.org/html/publications.htm.
26. International Society for Biological and Environmental Repositories (ISBER) (2005) Best practices for Repositories I: Collection, Storage and Retrieval of Human Biological Materials for Research. *Cell Pres Technol.* 3: 5–48.
27. Available by writing AATB Business Office, 1320 Old Chain Bridge Road, Suite 450, McLean VA, 22101.
28. Atkin, G., Daley, F.M., Bourne, S., Glynne-Jones, R., Northover, J., and Wilson GD. (2006) The effect of surgically induced ischaemia on gene expression in a colorectal cancer xenograft model. *Br. J. Cancer* 94, 121–127.

29. Lepor, H. and Kaci, L. (2003) Contemporary evaluation of operative parameters and complications related to open radical retropubic prostatectomy. *Urology* 62: 702–706.
30. Gaston, S.M., Soares, M.A., Siddiqui, M.M., Vu, D., Lee, J.M., Goldner, D.L., Brice, M.J., Shih, J.C., Upton, M.P., Perides, G., Baptista, J., Lavin, P.T., Bloch, B.N., Genega, E.M., Rubin, M.A., and Lenkinski, R.E. (2005) Tissue-print and print-phoresis as platform technologies for the molecular analysis of human surgical specimens: Mapping tumor invasion of the prostate capsule. *Nat. Med.* 11: 95–101.
31. Latham, G., Magotra, A., and Gaston, S.M. (2006) Cancer biomarker quantification using RNA extracted from tumor tissue print micropeels. *Ambion TechNotes* **13**(2): 25–26 www.ambion.com/techlib.
32. The Association of University Technology Managers http://www.autm.net.

# Index

Printed in the United States of America